NATIONAL SECURITY RESEARCH DIVISION

# Characterizing National Exposures to Infrastructure from Natural Disasters

## Data and Methods Documentation

Anu Narayanan, Henry H. Willis, Jordan R. Fischbach, Drake Warren, Edmundo Molina-Perez, Chuck Stelzner, Kathleen Loa, Lauren Kendrick, Paul Sorensen, Tom LaTourrette

Prepared for the U.S. Department of Homeland Security, Office of Infrastructure Protection

For more information on this publication, visit www.rand.org/t/rr1453z1

Library of Congress Cataloging-in-Publication Data
ISBN: 978-0-8330-9500-8

Published by the RAND Corporation, Santa Monica, Calif.


*Cover Image: © Fotolia/kentoh*

# Preface

The Department of Homeland Security, National Preparedness and Programs Directorate, Office of Infrastructure Protection, has asked the RAND Corporation to analyze exposures of national infrastructure systems to natural hazards and how these exposures are expected to evolve in response to climate change. RAND's analysis, which is documented in additional forthcoming reports, takes into account 11 hazards and five infrastructure sectors. This report describes the data and methods used by the RAND team to complete the analysis.

## The RAND Homeland Security and Defense Center

The research in this report was conducted in the Homeland Security and Defense Center (HSDC), which performs analysis to prepare and protect communities and critical infrastructure from natural disasters and terrorism. HSDC projects examine a wide range of risk-management problems, including coastal and border security, emergency preparedness and response, defense support to civil authorities, transportation security, domestic intelligence, and technology acquisition. HSDC clients include the U.S. Department of Homeland Security, the U.S. Department of Defense, the U.S. Department of Justice, and other organizations charged with security and disaster preparedness, response, and recovery.

HSDC is a joint center of two research divisions: RAND Justice, Infrastructure, and Environment and the RAND National Security Research Division. RAND Justice, Infrastructure, and Environment is dedicated to improving policy and decisionmaking in a wide range of policy domains, including civil and criminal justice, infrastructure protection and homeland security, transportation and energy policy, and environmental and natural resource policy. The RAND National Security Research Division conducts research and analysis for all national security sponsors other than the U.S. Air Force and the Army. The division includes the National Defense Research Institute, a federally funded research and development center whose sponsors include the Office of the Secretary of Defense, the Joint Staff, the Unified Combatant Commands, the defense agencies, and the U.S. Department of the Navy. The National Security Research Division also conducts research for the U.S. intelligence community and the ministries of defense of U.S. allies and partners.

Questions or comments about this report should be sent to the project leader, Henry Willis (hwillis@rand.org). For more information about HSDC, see www.rand.org/hsdc or contact the director at hsdc@rand.org.

# Contents

# Figures and Tables

## Figures

## Tables

# Summary

The United States relies on a number of infrastructure systems—roads, the electric grid, ports, telecommunications networks, refineries, and the like—for carrying out basic social and economic functions. Disruptions of these systems could impose potentially significant economic, social, environmental, and national security consequences. The U.S. Department of Homeland Security Office of Infrastructure Protection is charged with identifying and prioritizing strategies and investments for improving the resilience of specific infrastructure systems in specific regions. A necessary first step in fulfilling this role is understanding how infrastructure is exposed to natural disasters.

To support this work, the Department of Homeland Security, National Preparedness and Programs Directorate, Office of Infrastructure Protection, asked RAND to analyze national exposures to infrastructure from natural disasters. RAND's analysis includes 11 natural hazards and five infrastructure sectors. Tables S.1 and S.2 list the hazards, infrastructure sectors, and subsectors included in this analysis.

This report serves as the technical documentation and reference document for the data, methods, and analytic approach used for this study. The report also documents how each infrastructure type and hazard is represented in data sets to act as a reference for any use of the

**Table S.1**
**Hazards Included in Each Chapter**

| Chapter Two: Climate-Adjusted Hazards[a] | Chapter Three: Climate-Unadjusted Hazards |
|---|---|
| Coastal flooding | Earthquakes |
| Extreme temperature | Hurricane winds |
| Meteorological drought | Ice storms |
| Wildfires | Riverine flooding |
| | Tsunamis |
| | Tornadoes |
| | Landslides |

[a] *Climate-adjusted hazards* refer to those hazards for which we have projected future trends based on credible relationships between climate projections and hazard effects. For these hazards, this analysis enables both present-day and future views of infrastructure exposure. For hazards without such data, this analysis only enables a present-day view. When hazards are *climate-adjusted*, this simply means that credible projections are available for the effect of climate on these hazards and enable the user to look ahead to how infrastructure exposure might look in the future. So the "adjustment" refers to an ability to project into the future. Climate data for two hazards at a given point in time are never considered together unless those data are available for both hazards.

**Table S.2**
**Infrastructure Sectors and Subsectors Included in Analysis**

| Infrastructure Sector | Subsectors |
|---|---|
| Chemical industry | Chemical manufacturing facilities |
| Communications | Internet exchange points |
| Energy (including nuclear power) | Electric power generation plants |
| | Electric power substations |
| | Energy power transmission lines |
| | Energy distribution and control facilities |
| | Natural gas import/export points |
| | Natural gas processing plants |
| | Nuclear fuel facilities |
| | Nuclear power plants |
| | Oil and natural gas pipelines |
| | Petroleum, oil, and lubricants storage facilities |
| | Refineries |
| Transportation | Airports |
| | Canals |
| | DHS-identified railroad bridges |
| | Railroad stations |
| | Railroad transit lines |
| | DHS-identified railroad tunnels |
| | Railroad yards |
| | DHS-identified road bridges and tunnels |
| | Coastal, Great Lakes, and inland ports |
| | FAA air route traffic control centers |
| | Fixed-guideway transit systems, stations, and lines |
| | Intermodal terminal facilities |
| | Interstate highways |
| | Locks |
| Water supply and wastewater treatment | Dams[a] |
| | Wastewater treatment plants |

[a] Dams serve multiple purposes depending on the structure and can sometimes be grouped with water or transportation.

data. For each analyzed hazard, this report includes a brief background that describes potential infrastructure impacts, and relevant metrics; a list of sources used in compiling hazard data; and an overview of existing methods and applications or modifications used to analyze regional exposure to varying levels of hazard severity. When analyzing infrastructure exposures with these data, it is important to understand this information to ensure that the analysis results reflect the scope, precision, and completeness of the data. Failure to appropriately use the data could result in analysis that misrepresents exposures.

Appendix A provides an overview of all hazard and infrastructure data used to complete this analysis. Analytic findings about current and future exposures of infrastructure in the United States drawn from this data analysis are documented in a separate report.

# Acknowledgments

This work benefited from the constructive guidance and review received from many people during the completion of this study. In particular, the study team is grateful for the insights provided by Lisa Barr, Sarah Ellis-Peed, Debra Knopman, Bob Kolasky, Meredith Secor, and Kate White. We also thank David Groves, Seth Guikema, Lucia Schmit, and Julie Waters for providing extremely thorough reviews of the document, which greatly improved the quality of the work.

## Abbreviations

| | |
|---|---|
| AEP | annual exceedance probability |
| BFE | base flood elevation |
| CDD | consecutive dry days |
| CMI | Crop Moisture Index |
| CMIP5 | Coupled Model Intercomparison Project Phase 5 |
| CONUS | Continental United States |
| DC | Digital Coast |
| DEM | digital elevation model |
| DHS | U.S. Department of Homeland Security |
| FAA | Federal Aviation Administration |
| FEMA | Federal Emergency Management Agency |
| FHZ | Flood Hazard Zone |
| FIRM | Flood Insurance Rate Map |
| FS | flood stage |
| FSim | Large Fire Simulator |
| GCM | General Circulation Model |
| GEV | Generalized Extreme Value |
| HEC-HMS | Hydrologic Engineering Center Modeling System |
| HSDC | Homeland Security and Defense Center |
| HSIP | Homeland Security Infrastructure Program |
| IPCC | Intergovernmental Panel on Climate Change |
| IPCC, AR4 | Intergovernmental Panel on Climate Change, Fourth Annual Report |
| IPCC, AR5 | Intergovernmental Panel on Climate Change, Fifth Annual Report |
| KBDI | Keetch-Byram Drought Index |
| NED | National Elevation Dataset |
| NFDRS | National Fire Danger Rating System |
| NFIP | National Flood Insurance Program |
| NOAA | National Oceanic and Atmospheric Administration |
| NPCC | New York Panel on Climate Change |

| | |
|---|---|
| NWS | National Weather Service |
| PDSI | Palmer Drought Severity Index |
| RCP | Representative Concentration Pathways |
| RMS | Risk Management Solutions |
| RSLR | relative sea level rise |
| SFHA | special flood hazard areas |
| SLR | sea level rise |
| SPIA | Sperry-Piltz Ice Accumulation |
| SREX | Special Report on Managing Risks of Extreme Events |
| USACE | U.S. Army Corps of Engineers |
| USGS | U.S. Geological Survey |
| WFP | Wildland Fire Potential |

CHAPTER ONE

# Introduction

The United States relies on a number of infrastructure systems—roads, the electric grid, ports, telecommunications networks, refineries, and the like—for carrying out basic social and economic functions. Disruption of these systems could impose potentially significant economic consequences in the form of delays and increased costs. To minimize such economic losses, it is important that critical infrastructure systems be resilient—that is, able to maintain and regain operational capacity in the aftermath of a major disruption, and to recover relatively quickly and at low cost following the disruption.

In recent years, debates on federal infrastructure investment policies have focused largely on the need to renew aging facilities nearing the end of their design life as well as to add new capacity to meet increased demand. With the severity of the recent economic recession, there has also been strong interest in the role of infrastructure investment as a near-term economic stimulus, and as a way to create the foundations for longer-term economic growth. As shown by the significant damages and lasting disruptions caused by Hurricanes Katrina and Sandy, however, there can also be a significant federal interest in policies aimed at enhancing the resilience of the nation's critical infrastructure.

The U.S. Department of Homeland Security (DHS) Office of Infrastructure Protection is charged with identifying and prioritizing strategies and investments for improving the resilience of specific infrastructure systems in specific regions. A necessary step in fulfilling this role is understanding how infrastructure is exposed to natural disasters.

To support this work, the DHS, National Preparedness and Programs Directorate (NPPD), Office of Infrastructure Protection asked RAND to analyze national exposures to infrastructure from natural disasters. The objective of the completed analysis is to help NPPD highlight patterns of exposure that exist both today and in the future when looking across multiple hazards and infrastructures. For instance, this analysis could help answer such questions as: Which areas of the country are currently exposed to more-intense or multiple hazards? How is the picture likely to change in the future? How might climate change alter exposure patterns? The main focus of this analysis is not on definitively answering all of these questions but rather to provide NPPD and DHS a way to begin to screen for regions and infrastructures within the United States that might be more susceptible to natural hazards today and in the future.

Infrastructure data used in this analysis are drawn from the Homeland Infrastructure Foundation-Level Data Homeland Security Infrastructure Protection (HSIP) Gold data set.[1]

[1] For more information, see the Homeland Infrastructure Foundation-Level Data Subcommittee Online Community, undated.

Because this data set contains proprietary commercial data sets, this analysis describes expo sure to infrastructure aggregated at the county level so as not to reveal any of the underlying proprietary databases. Because of the uncertainty in natural disaster events, natural hazards are described with measures of intensity and likelihood: for example, sustained wind speeds with a 100-year return interval or flood depths with a 500-year return interval, or peak ground acceleration from seismic activity with a 2,500-year return interval. The hazard data used in this analysis include information of this type compiled from a variety of sources described in the rest of the report.

The analytical steps involved in completing this work included:

- Compiling present and future hazard data (from different sources, described in Chapters Two and Three)
- Identifying infrastructure asset types to include in the analysis (using HSIP Gold as the only source of infrastructure asset data)
- Deciding whether a given asset is truly at risk of "exposure" to a given hazard based on inherent attributes of each asset,[2] as well as whether exposure of the asset to a particular hazard is likely to have consequences. For instance, just existing in a region that is prone to wildfires does not actually expose an underground electrical distribution line to a wildfire.
- Overlaying infrastructure and hazard data to provide views of infrastructure exposure to one or more hazards, now and in the future

We note that the completed work is not a risk assessment, and focus solely on assessing whether different infrastructures are likely to be exposed to different hazard types. Conducting a complete risk assessment would require vulnerability and impact analyses for each considered infrastructure and hazard type. RAND's analysis considers 11 natural hazards and five infrastructure sectors. Table 1.1 lists the hazards included in each chapter, Table 1.2 lists the return periods considered for each hazard,[3] and Table 1.3 lists the infrastructure sectors and subsectors included in RAND's analysis. We also note that of the several available data sets and methods for analyzing natural hazards, we necessarily had to choose one to carry out this analysis. Choosing one way to approach the analysis for a given hazard certainly introduces uncertainty in which elements of infrastructure are vulnerable to the different hazards. Given that the objective of this work is to provide a first look at infrastructure exposure to natural hazards from a screening perspective, we believe that the embedded uncertainties do not take away from the usefulness of the analysis.

---

[2] See Chapter Four for a description of the dimensions used to decide whether an asset type would be included in this analysis.

[3] The differences in return periods reflect the variability in the types of data we were able to acquire for the different hazards. Other dimensions of the data used for each hazard are listed in a table in the corresponding section of this report.

**Table 1.1**
**Hazards Included in Each Chapter**

| Chapter Two: Climate-Adjusted Hazards[a] | Chapter Three: Climate-Unadjusted Hazards |
|---|---|
| Coastal flooding | Earthquakes |
| Extreme temperature | Hurricane winds |
| Meteorological drought | Ice storms |
| Wildfires | Riverine flooding |
| | Tsunamis |
| | Tornadoes |
| | Landslides |

[a] *Climate-adjusted hazards* refer to those hazards for which we have projected future trends based on credible relationships between climate projections and hazard effects. For these hazards, this analysis enables both present-day and future views of infrastructure exposure. For hazards without such data, this analysis only enables a present-day view. When hazards are *climate-adjusted*, this simply means that credible projections are available for the effect of climate on these hazards and enable the user to look ahead to how infrastructure exposure might look in the future. So the "adjustment" refers to an ability to project into the future. Climate data for two hazards at a given point in time are never considered together unless those data are available for both hazards.

**Table 1.2**
**Return Periods Associated with Each Hazard**

| Hazard | Return Periods |
|---|---|
| Coastal flooding | 2/5/10/20/50/100 years |
| Extreme temperature | 2/5/10/20/50/100 years |
| Drought | 75th/95th percentile KBDI |
| Wildfires | N/A |
| Earthquakes | 500 and 2,500 years |
| Hurricane winds | 10/20/50/100/200/500/1,000 years |
| Ice storms | 50 years |
| Riverine flooding | 100 years |
| Tsunamis | ≤500 years |
| Tornadoes | 100,000 years |
| Landslides | N/A |

NOTE: KBDI = Keetch-Byram Drought Index. N/A = not available.

The report documents how each infrastructure type and hazard is represented in data sets to act as a reference for any use of the data. For each analyzed hazard, this report includes a brief background that describes potential infrastructure impacts and relevant metrics; a list of sources used in compiling hazard data; and an overview of existing methods and applications or modifications used to analyze regional exposure to varying levels of hazard severity. When

**Table 1.3**
**Infrastructure Sectors and Subsectors Included in Analysis**

| Infrastructure Sector | Subsectors |
|---|---|
| Chemical industry | Chemical manufacturing facilities |
| Communications | Internet exchange points |
| Energy (including nuclear power) | Electric power generation plants |
| | Electric power substations |
| | Energy power transmission lines |
| | Energy distribution and control facilities |
| | Natural gas import/export points |
| | Natural gas processing plants |
| | Nuclear fuel facilities |
| | Nuclear power plants |
| | Oil and natural gas pipelines |
| | Petroleum, oil, and lubricants storage facilities |
| | Refineries |
| Transportation | Airports |
| | Canals |
| | DHS-identified railroad bridges |
| | Railroad stations |
| | Railroad transit lines |
| | DHS-identified railroad tunnels |
| | Railroad yards |
| | DHS-identified road bridges and tunnels |
| | Coastal, Great Lakes, and inland ports |
| | FAA air route traffic control centers |
| | Fixed-guideway transit systems, stations, and lines |
| | Intermodal terminal facilities |
| | Interstate highways |
| | Locks |
| Water supply and wastewater treatment | Dams[a] |
| | Wastewater treatment plants |

[a] Dams serve multiple purposes depending on the structure and can sometimes be grouped with water or transportation.

analyzing infrastructure exposures with this data, it is important to understand this information to ensure that the results of analysis reflect the scope, precision, and completeness of the data. Failure to appropriately use the data could result in analysis that misrepresents exposures.

The analysis addresses infrastructure exposure to hazards in the continental United States, excluding Alaska, Hawaii, and other territories. Hazards are adjusted to reflect climate change when credible relationships between climate projections and hazard effects exist.[4] These climate-adjusted hazards are described in Chapter Two. In the absence of such data and methods, hazards are treated as unchanging with climate change. Hazards not adjusted for climate change are described in Chapter Three.

Adjustments to hazards reflect changes in frequency, severity, or impact area due to climate change. We include climate-adjusted hazards when possible in the interest of accounting for future climate change where possible. The lowest common denominator for all hazards is the present day snapshot, where no hazards are adjusted for future climate change. One limitation of this approach is that when looking ahead at cross-hazard interactions, some hazards would be climate-adjusted and others would not. This does not mean in all cases that we think these climate-unadjusted hazards are insensitive to climate change. This simply means we were unable to capture how climate change would affect these hazards.

For several of these hazards, possible climate change interactions are beyond those captured in this analysis. Changes in the intensity or frequency of hurricanes could change the distribution of hurricane-force winds along coasts and modify risks of coastal flooding beyond the effects of sea level rise (SLR). Changes in precipitation patterns could modify exposures to riverine flooding, landslides, and ice storms. To the extent these effects amplify hazards, analysis using the data described in this report underestimates natural hazard exposures. To the extent these effects diminish hazards, such analysis overestimates natural hazard exposures.

The report concludes by describing the process used to identify the specific infrastructure sectors to include in the analysis and by describing the approach used to assess interactions between infrastructure and hazards.

Analytic findings about current and future exposures of infrastructure in the United States drawn from analysis of the data are documented in a separate report (Willis et al., 2016).

### 1.1. References

Homeland Infrastructure Foundation-Level Data Subcommittee Online Community, "Welcome to the HIFLD Subcommittee Home Page," web page, undated. As of April 7, 2016:
https://gii.dhs.gov/hifld

Willis, Henry H., Anu Narayanan, Jordan Fischbach, Edmundo Molina-Perez, Chuck Stelzner, Katie Loa, and Lauren Kendrick, *Current and Future Exposure of Infrastructure in the United States to Natural Hazards*, Santa Monica, Calif.: RAND Corporation, RR-1453-DHS, 2016.

---

4 We searched for existing literature, methods, and data that could be used to describe the future probabilistic exposure or model the effect of climate change on the current probabilistic exposure. Where we found existing data and methods, we used them. Where we did not, we did not reflect climate change.

CHAPTER TWO

# Climate-Adjusted Hazards

This chapter describes the data and methods used to analyze the following four climate-adjusted hazards:

- coastal flooding
- extreme temperature
- meteorological drought
- wildfires.

For each hazard, we provide a brief overview of the significance of the hazard—both at present and in the future with climate change—and describe the data sets and analysis methods used to assess the likelihood of regional exposure to varying levels of hazard severity. Where future hazard projections are presented, data reflect hazards in three time periods: 2040, 2065, and 2100. The first two of these periods represent common time horizons for infrastructure planning. The third represents a longer time horizon for which climate scenarios diverge significantly.

Two different types of climate projections are used to project future changes in these hazards as a response to climate change. For the drought, wildfire, and extreme temperature hazards, the Coupled Model Intercomparison Project Phase 5 (CMIP5) climate projections are used for analysis. For the tidal flooding, storm surge, and permanent inundation, all grouped broadly under the coastal flooding hazard, National Oceanic and Atmospheric Administration (NOAA) year 2100 SLR projections were used as inputs.

We combined these projections into a set of "aggregated climate change scenarios" that could be used across all four climate-adjusted hazards. These aggregated scenarios are listed in Table 2.1 and described in two dimensions: (1) emissions level and (2) percentile of the general circulation models (GCMs) considered. CMIP5 projections are classified in two dimensions: (1) representative concentration pathways (RCP) and (2) percentile of the GCMs considered. We first estimated the projected SLR values associated with each CMIP5 RCP and mapped these SLR values to the 2100 NOAA SLR scenarios: low SLR (low), low-to-intermediate SLR (int-low), intermediate-to-high SLR (int-high), and high SLR (high). Comparing the derived CMIP5 SLR values with the NOAA SLR values and using the model percentile information for each CMIP5 projection enabled us to create five aggregated climate scenarios: (1) Low Emissions–Median, (2) Low Emissions–High, (3) High Emissions–Median, (4) High Emissions–High, and (5) Worst Case, which are mapped across both types of climate projections.

**Table 2.1**
**Climate Scenario Mapping**

| Aggregated Climate Scenario | RCP | Percentile of the 22 GCMs Considered | IPCC AR5 SLR Value 2100 (approx. ft.) | Suggested Mapping to NOAA/NCA SLR | NOAA/NCA SLR Value 2100 (approx. ft.) |
|---|---|---|---|---|---|
| Low Emissions–Median | RCP 4.5 | 50 | 1.7 | Low | 1.44 |
| Low Emissions–High | RCP 4.5 | 95 | 2.3 | Int-Low | 2.43 |
| High Emissions–Median | RCP 8.5 | 50 | 2.4 | Int-Low | 2.43 |
| High Emissions–High | RCP 8.5 | 95 | 3.2 | Int-High | 4.63 |
| Worst Case[a] | RCP 8.5 | Max | N/A | High | 7.12 |

NOTE: IPCC, AR5 = Intergovernmental Panel on Climate Change, Fifth Annual Report.

NCA = National Climate Assessment.

[a] Worst Case refers to the worst among our aggregated climate scenarios and not the absolute worst possible future.

For instance, the aggregate Low Emissions–Median climate scenario corresponds in the CMIP5 projections to emissions scenario RCP 4.5 and the conditions described by the 50th percentile of the 22 GCMs considered. Table 2.2 lists these 22 GCM models, with the names denoting the author modeling group and the ensemble used for the projections. These models include the RCP 4.5 and 8.5 emissions scenarios and cover the 2006–2100 time period.[1]

**Table 2.2**
**Climate Models Used in This Analysis**

| Climate Models |
|---|
| ACCESS1.0_r1i1p1 |
| GFDL.ESM2G_r1i1p1 |
| bcc.csm1.1_r1i1p1 |
| CESM1.BCG_r1i1p1 |
| CCSM4_r2i1p1 |
| CanESM2_r2i1p1 |
| CSIRO.Mk3.6.0_r1i1p1 |
| MIROC.ESM_r1i1p1 |
| inmcm4_r1i1p1 |
| IPSL.CM5A.LR_r1i1p1 |
| NorESM1.M_r1i1p1 |
| CanESM2_r1i1p1 |

[1] For extreme temperature, data were only available on the CMIP5 data set for 21 of the 22 models. Additionally, one of the models had no data for the time period 2100, so only 20 of the 22 models were run.

**Table 2.2—Continued**

| Climate Models |
|---|
| CCSM4_r1i1p1 |
| GFDL.ESM2M_r1i1p1 |
| IPSL.CM5A.LR_r2i1p1 |
| IPSL.CM5A.MR_r1i1p1 |
| MIROC5_r1i1p1 |
| MIROC5_r2i1p1 |
| MIROC.ESM.CHEM_r1i1p1 |
| MPI.ESM.LR_r1i1p1 |
| MPI.ESM.MR_r1i1p1 |
| MRI.CGCM3_r1i1p1 |

NOTE: Data were not available for analysis of extreme temperatures using the MIROC5_r2i1p1 model for any time periods or for the bcc.csm1.1_r1i1p1 model in the 2100 time period.

## 2.1. Coastal Flooding

Coastal communities in the United States are exposed to a range of potential hazards, but the most concerning and damaging of these events are coastal floods caused by high tides and extreme storms. The latter—including both tropical depressions and extra-tropical storms that threaten the Eastern Seaboard and Gulf Coast of the United States—can produce storm surges and waves that can damage or destroy coastal property, threaten the health and wellbeing of residents, and degrade or destroy natural habitats (Burkett and Davidson, 2012).

Natural and anthropogenic factors influence the frequency and severity of flooding from tidal flooding and coastal storms (Woodruff, Irish, and Camargo, 2013). For instance, poorly understood annual, interannual, and decadal natural climate variations influence the frequency and intensity of coastal storms, and there is a limited observed historical record from which to improve scientific understanding and make predictions.

A warming climate can further exacerbate the risks posed by coastal flooding, and also adds additional uncertainty to flood risk projections. Climate change is expected to increase the risk of flooding to coastal communities and assets through several pathways. Most significant among these is SLR, which can impact tidal range, storm surge, and wave heights (Karl et al., 2008). In turn, SLR directly increases the frequency and severity of coastal flooding by increasing the depth of flooding when high tides and coastal storms occur. Global mean sea levels have risen approximately eight inches since 1900 (Church and White, 2011), which have exacerbated the damage caused by storms like Hurricane Sandy (Leifert, 2015). In some cases, SLR in the future could lead to land area becoming permanently inundated unless steps are taken to protect communities from flooding.

Climate factors influencing global and local sea level changes include thermal expansion of ocean water, melting ice sheets, wind, atmospheric pressure, oceanic circulation variances, and water density differences. But projecting future SLR is challenging because of the uncertain rate of atmospheric warming and uncertainty in regard to the underlying physical processes. Many previous studies, including the Intergovernmental Panel on Climate Change, Fourth Annual Report (IPCC AR4), have assumed that thermal expansion is the dominant contributor to SLR (Pachauri and Reisinger, 2007). But future SLR could be driven by the rate and magnitude of ice sheet melting from Greenland and West and East Antarctica. Since the full range of ice sheet melting scenarios were not considered, some scientists argue that the IPCC AR4 failed to estimate the full range of SLR possibilities (Horton et al., 2008). The Intergovernmental Panel on Climate Change, Fifth Annual Report (IPCC AR5), includes SLR projections based on ice sheet and ocean models, and as a result raises the 2081–2100 global sea level projection compared with the IPCC AR4 (Miller et al., 2013).

For the Third U.S. National Climate Assessment (Walsh et al., 2014), NOAA developed four global mean SLR scenarios by 2100, which we adopt in this study (Parris et al., 2012). These scenarios are shown in Figure 2.1, and summarized as follows:

- **Lowest:** derived from a linear extrapolation of the historical SLR rates from tide records.
- **Intermediate-Low:** derived from upper end of the IPCC AR4 global SLR projection using the B1 emissions scenario.
- **Intermediate-High:** derived from the average of high-end semi-empirical, global SLR projections, which include recent ice sheet loss.
- **Highest:** derived from IPCC AR4 global SLR projections and maximum possible glacier and ice sheet loss.

**Figure 2.1**
**Global Mean SLR Scenarios for the Third National Climate Assessment**

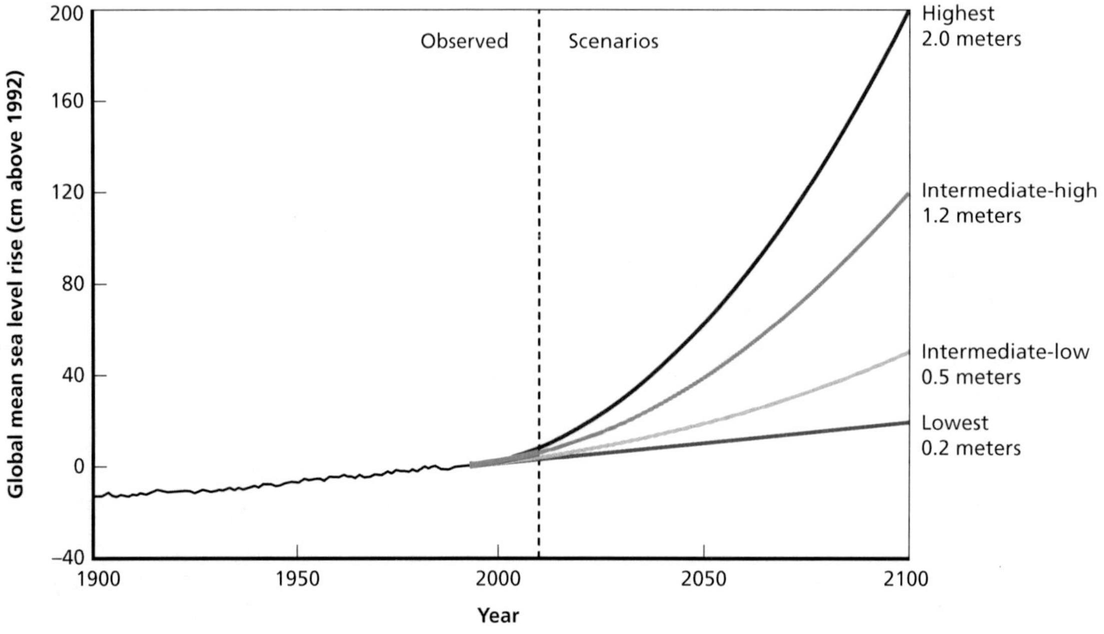


SOURCE: Parris et al., 2012.
RAND *RR1453/1-2.1*

SLR can be further exacerbated by localized factors, including land sinking or subsidence. When land subsides, the effect sums with eustatic SLR to create a relative sea level rise (RSLR) rate, which includes both processes. Subsidence rates vary across different U.S. coastal regions, with notably higher rates of subsidence in coastal Louisiana and the mid-Atlantic.

Figure 2.2 shows relative sea level change in 2013 for tide gauge stations across U.S. coastal areas. Note that in some areas of the Pacific Coast and Canada, RSLR is decreasing—this is due to land uplift in these areas that is sufficient to offset global eustatic SLR.

In addition to SLR, there is a significant amount of literature describing the potential links between climate change and future increases in the frequency or severity of tropical depressions. These effects could modify exposures to coastal flooding and extreme winds (see the hurricane winds discussion later in this report). Though there is growing consensus that changes in climate will have effects on the intensity and frequency of storms, the literature does not describe how the distribution of exposure will be modified. Thus, these effects were not included in the climate scenarios described here. Depending on where exposures are occurring, the resulting analysis either under- or overestimates exposures.

**Figure 2.2**
**RSLR at NOAA Tide Gauge Stations**

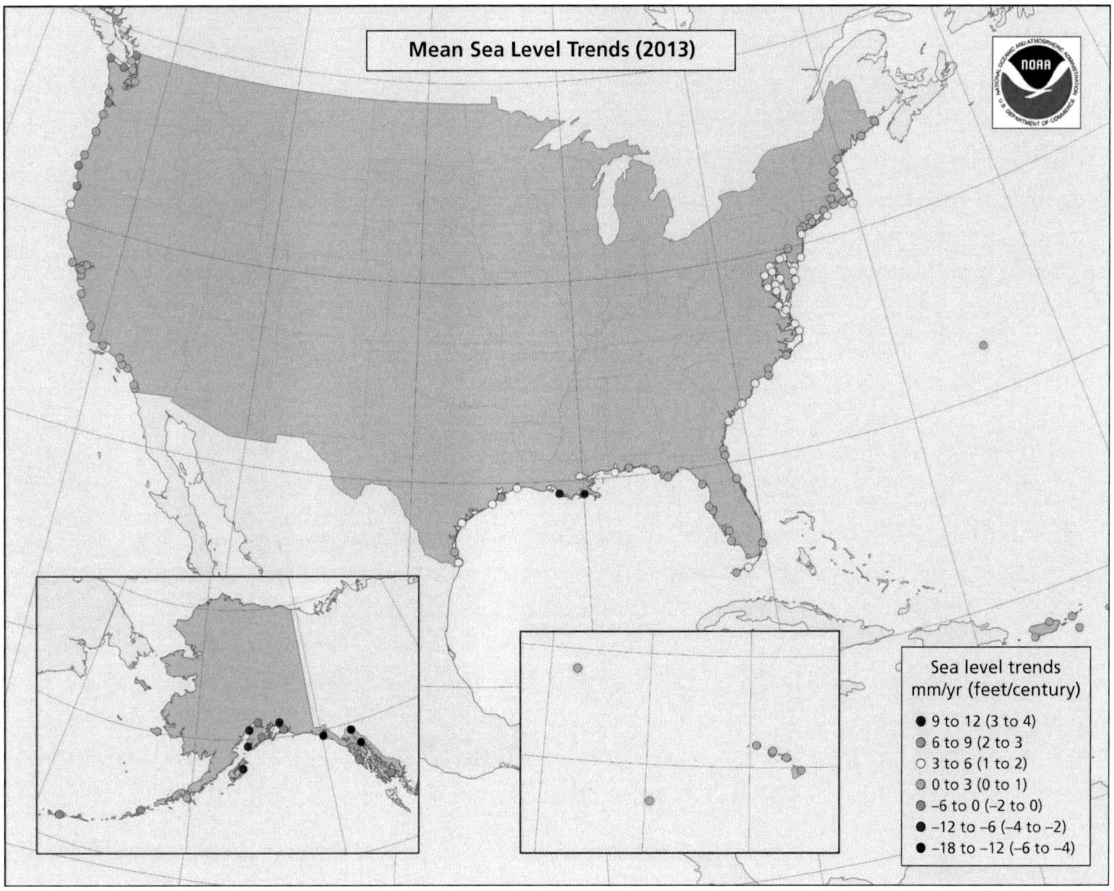


RAND *RR1453/1-2.2*

SOURCE: NOAA/National Ocean Service (2013a).
NOTE: Figure shows the location of each gauge station, along with an estimate of the sea level trend observed as of 2013.

The primary metric for coastal flood hazard is expected flood depth (in feet), estimated at different annual exceedance probabilities (AEPs). The flood exposure analysis determines the flood depth for each facility location (infrastructure) or county centroid (population), and uses a simple binary assessment to determine if the depth exceeds a critical threshold. Critical thresholds were set at 1 foot and 6 feet of flood depth, representing low and high damage intensities, respectively. This exercise was repeated for all three coastal flood hazard types described below.

### 2.1.1. Data Sources

The coastal flood exposure analysis draws on existing federal data sources developed to support flood risk and climate resilience planning. The methods to incorporate these data are described in the next subsection.

Key coastal flooding data sources for this analysis include:

- **Digital elevation model:** We adapted a digital elevation model (DEM) from the U.S. Geological Survey (USGS) National Elevation Dataset (NED) (USGS, 2014) at a 1 arc-second resolution. These ground elevations were used to calculate flood depths for each facility or county centroid.
- **Coastal land subsidence rates:** The source for coastal land subsidence rates is observed tidal gauge data, provided via NOAA's Regional Mean Sea Level Trends data set, which covers the period from 1920 to 2006 (NOAA/National Ocean Service, 2013b). These data report sea level trends at NOAA coastal monitoring stations across the country, and they were used to identify locally varying land subsidence rates. These trends are adopted as-is and projected forward; no scenario adjustment is planned for these data due to lack of sources that provide a systematic nationwide assessment of future scenarios or changes.
- **Permanent inundation SLR scenarios:** SLR scenarios, ranging from 1 to 6 feet above current conditions, were also adapted from the NOAA Digital Coast (DC) analysis (2015).
- **NOAA global mean SLR scenarios:** We adopted four consensus global SLR scenarios developed by NOAA to support the Third National Climate Assessment (Parris et al., 2012; Burkett and Davidson, 2012). These scenarios reflect different assumptions about the future SLR acceleration rate (see above).
- **Tidal flood projections with SLR:** As discussed below, Kriebel and Geiman (2013) developed a method to project tidal flood stages (FSs) at a range of exceedance probabilities with SLR incorporated. These data were obtained from the U.S. Army Corps of Engineers (USACE) and are applied for the higher-frequency flood analysis.
- **Federal Emergency Management Agency (FEMA) base flood elevations:** The primary data source for current low-frequency coastal flood risk is FEMA's National Flood Hazard Layer, which provides FEMA's estimate of the 1 percent AEP flood elevation. The 1 percent AEP elevation is used to identify special flood hazard areas (SFHA) for the National Flood Insurance Program's (NFIP) Flood Insurance Rate Maps (FIRMs). Residents of the SFHA are required to purchase flood insurance via NFIP under most circumstances—including, for example, in order to qualify for a federally insured mortgage. The National Flood Hazard Layer includes flood maps that have been formally adopted and digitized. The actual flood elevation and flood depth data do not include all areas exposed to flood risk in the United States, however, with gaps from (1) paper maps that

have not yet been digitized and (2) a slow analytic update and approval process for new maps. As a result, not all coastal regions can be included in the analysis focused on the 100-year floodplain.[2] SLR and subsidence were added subsequently to FEMA's 100-year elevations according to the methods described in the following section.

### 2.1.2. Analysis Methods

We divide the hazard analysis described in this section into three parts, roughly corresponding to differing likelihood ranges for coastal flooding. The first portion considers the possibility of permanent inundation due to future SLR. The second focuses on tidal or "nuisance" flooding, which occurs on a more-frequent basis (e.g., less-than-one-year, one-year, or two-year event) but typically produces only minor damage during each event (Spanger-Siegfried, Fitzpatrick, and Dahl, 2014). The final subsection considers flooding from extreme storm events and tropical depressions, which have lower annual likelihoods (e.g., 100-year event) but can cause extensive damage and disruption. The frequency and damage from all three coastal flood types will be influenced by future SLR.

#### *SLR and Permanent Inundation*

The first hazard approach for coastal flooding was to consider areas that would be permanently inundated by SLR in future climate scenarios. To estimate flood depths in this case, we adapted the NOAA DC analysis to directly consider the effects of a fixed level of SLR ranging from 1 foot to 6 feet above the current vertical datum (NAVD88). As DC data include flood depth DEMs for each of these SLR scenarios, the flood depth DEMs were left as-is to be run against relevant infrastructure. Specifically, flood elevations were compared with a 10-meter resolution DEM for each of the six cases, creating a depth layer for each case. These depth scenarios were later matched to the nearest NOAA SLR scenario at different points in time (see below) to develop a common set of future scenarios, though the DC analysis uses fixed thresholds at each foot and does not consider alternate pathways for SLR acceleration over time. Table 2.3 summarizes the data sources, scenarios, and procedure used to estimate flood depths from permanent inundation. The years 2040 and 2065 represent scenario projections 25 years and 50 years into the future, which were intervals of interest for DHS. The year 2100 is a commonly used end date for global and national climate projections, and is the furthest date for which downscaled future climate projections are typically available.

#### *Tidal Inundation and Higher-Frequency Coastal Flood Events*

Historical coastal tide levels and flood data are collected from gauge stations on the Pacific, Atlantic, and Gulf Coasts (Figure 2.2). The short lengths of the historical record and distance between gauge stations, however, limit the data available from these stations. In addition, quantitative comparison between simulated extreme events and station data is often difficult because of variations in scale and trends (Osborn and Hulme, 1997).

To address these limitations, the Kriebel method describes an approach to determine major FS when no National Weather Service (NWS) FS data are available at these gauges. Using a generic FS allows for a reference comparison between different coastal areas within a given geographic region, with results that are consistent with moderate and major NWS

[2] As one notable example, updated flood risk information currently is not available from FEMA for coastal Louisiana.

**Table 2.3**
**Estimation of Coastal Flood Risk from Future Permanent Inundation**

| Analysis Factor | Description |
|---|---|
| Years | 2040, 2065, 2100 |
| Data | Flood heights:<br>NOAA DC water elevations |
| Scenarios | 1 foot to 6 feet of sea level above current mean higher high water, subsequently matched to the nearest NOAA SLR scenario in each time period:<br>• Lowest<br>• Intermediate-Low<br>• Intermediate-High<br>• Highest |
| Return periods | N/A |
| Models | N/A |
| Procedure | Calculate depth of flooding with each additional foot of SLR from regional DEMs |

FS data. The Kriebel method uses observed monthly extreme water levels from NOAA tide gauges: A major storm is aligned with extreme high tide when it is within three standard deviations of the mean, and moderate storm is aligned with extreme high tide when it is within two standard deviations of the mean (Kriebel and Geiman, 2013).

Essentially, the Kriebel method provides a simplified way to estimate tidal flood elevations at different exceedance probabilities in a consistent way for all NOAA tide gauge stations. SLR scenarios can then be added to these estimates in order to project future tidal flooding. Data were provided for exceedances ranging from the two-year to the 100-year event, but we generally consider the estimates from tidal flooding of this kind—which typically do not consider or include extreme storm surge events—to be more reliable in the two-year to the 20-year interval range.

For this analysis, we directly adopted data produced by Kriebel and Geiman (2013) using this method described above and provided to the research team by the USACE Institute for Water Resources. The data set included tidal flood elevations ranging from the two-year to 100-year event at each NOAA tide gauge station. We estimated RSLR by combining the NOAA global mean SLR scenarios (Figure 2.1) with local subsidence rates at each of these station locations, and then estimating the final RSLR for each year and scenario. These values were then summed with the Kriebel tidal flood elevations to produce scenario- and year-specific values for each gauge station and exceedance probability. Finally, we interpolated these values for points in-between the gauge stations, creating a data set filling the continental United States (CONUS) coastal counties.

After interpolating, we used DC DEMs with USGS NED DEMs to fill CONUS coastal counties. The DC DEMs were coarsened to match the 1-arc-second resolution of the NED, and the two data sets were then combined, retaining the maximum ground elevation in areas of overlap. We used this combined DEM to estimate a tidal flood depth for each location of interest in the analysis, repeated for each exceedance probability and RSLR scenario.

Table 2.4 summarizes the data sources, scenarios, and procedure used to estimate flood depths from tidal flooding.

**Table 2.4**
**Estimation of Coastal Flood Risk from Higher-Frequency Tidal Flood Events**

| Analysis Factor | Description |
|---|---|
| Years | 2015 (current conditions)<br>2040 (25 years in the future)<br>2065 (50 years in the future)<br>2100 (end of the century) |
| Data | **Flood elevations**<br>Kriebel tidal flood estimates (Kriebel and Geiman, 2013)<br><br>**Ground elevation**<br>NOAA DC ground elevations coarsened to 30-meter DEM<br>USGS NED<br>30-meter DEM<br><br>**Land subsidence rates**<br>NOAA regional mean sea level trends (1920–2006) |
| Scenarios | NOAA global mean SLR scenarios, combined with local observed land subsidence rates to create RSLR scenarios:<br>• Lowest<br>• Intermediate-Low<br>• Intermediate-High<br>• Highest |
| Return periods | 2/5/10/20/50/100 years |
| Models | N/A |
| Procedure | For each exceedance probability and RSLR scenario:<br>1. Estimate RSLR (SLR + subsidence) at each gauge station<br>2. Sum RSLR with Kriebel tidal flood estimate for each exceedance probability<br>3. Interpolate tidal flood scenarios for areas in between gauge stations<br>4. Combine DC and DEM data to create overall DEM<br>5. Subtract ground elevation (DEM) from flood elevation to estimate depth at each location of interest. |

#### *Low-Frequency Storm Surge Events*

Lastly, we developed coastal flood scenarios for low-frequency storm surge events (Table 2.5). This analysis builds on the most recent 100-year base flood elevations (BFEs) provided by FEMA as part of the National Flood Hazard Layer. As discussed above, these were only provided for certain coastal regions, so this analysis does not provide a complete picture of low-frequency flooding in CONUS.

To estimate future flood depths under different SLR scenarios, we first made a simplifying assumption that storm surge and wave flooding sum linearly with RSLR. This likely underestimates future flood depths in some areas, because it does not account for nonlinear effects from RSLR on storm surge propagation (e.g., RSLR leading to loss of coastal marsh that previously absorbed storm surge and wave energy). Nevertheless, such an assumption is necessary to conduct an assessment at this scale and level of resolution.

For this analysis, FEMA Flood Hazard Zones (FHZs) were divided into coastal and noncoastal (riverine) areas. Investigation of the literature revealed that a previous study used FHZ classifications for this purpose, taking FHZs of type V as a distinguishing feature of coastal

**Table 2.5**
**Estimation of Coastal Flood Damage from Low-Frequency Events**

| Analysis Factor | Description |
|---|---|
| Years | 2015 (current conditions)<br>2040 (25 years in the future)<br>2065 (50 years in the future)<br>2100 (end of the century) |
| Data | **Flood depths**<br>FEMA National Flood Hazard Layer coastal base flood elevations<br><br>**Ground elevation**<br>NOAA DC ground elevations coarsened to 30-meter DEM<br>USGS NED<br>30-meter DEM<br><br>**Land subsidence rates**<br>NOAA regional mean sea level trends (1920–2006) |
| Scenarios | NOAA global mean SLR scenarios, combined with local observed land subsidence rates to create RSLR scenarios:<br>• Lowest<br>• Intermediate-Low<br>• Intermediate-High<br>• Highest |
| Return periods | 2/5/10/20/50/100 years |
| Models | N/A |
| Procedure | Baseline<br>• Using ArcGIS, calculate flood depths for all FHZs with BFEs by subtracting ground elevation from FEMA flood elevations.<br><br>Climate projections (SLR)<br>• Calculate RSLR = SLR + subsidence for selected time slices<br>• Add RSLR to FEMA BFEs<br>• Calculate extent of future floodplain and resulting depths using "bathtub" approach. |

areas (Crowell et al., 2010).[3] Therefore, we retained all FHZs of type V as coastal, retaining additional adjacent FHZs of type A as determined by a local inspection of inland FHZ extent. Once complete, all noncoastal FHZs were removed from the analysis.

FEMA BFE lines represent absolute flood depth for a 100-year flood. As delivered, the BFEs contained depths in a number of different formats, each measured with respect to a different model of sea level. After converting the disparate depths into a single sea level model, the BFEs were used to dice the coastal FHZs into smaller areas. Each new FHZ area was assigned the maximum flood depth of the BFE lines it touched.

To complete the analysis, a modified "bathtub" approach was used. For each RSLR scenario, the scenario's relative flood depths were added to the BFE absolute flood depths. The summed flood depths at the outer edges of the FHZs were then extrapolated outward, and surrounding ground topography, using the same DEM as in the Kriebel method, was sub-

[3] FEMA classifies SFHAs into several zones, of which Zone V marks an area inundated by 100-year flooding with velocity hazard.

tracted in the areas outside of the FHZs, resulting in relative flood depths. All areas outside of and directly adjacent to the FHZs, which contained positive relative flood depths, were then retained, adding to the FHZs to create an extended FHZ area. Finally, a script added any outlying areas with positive relative flooding that were within one raster cell of the extended FHZ area, building the extended FHZ area out until all relevant additional areas had been added, yielding the final flooded area.

This analysis makes use of the best-available national data from federal agencies for coastal flood risk, both in current conditions and projected forward into a climate-altered future. But notable gaps remain. First, higher-resolution, localized SLR scenario projections might be needed for some areas of the country, because local conditions could lead to sea levels well above the global mean projections. For example, the New York Panel on Climate Change (NPCC) recently noted that the average rate SLR over the last century in New York City has been twice the rate observed globally, leading to a 1-foot increase in mean sea level for this area. NPCC also developed new SLR projections out to 2100 for New York and the mid-Atlantic region that are higher than the consensus IPCC AR5 scenarios (Horton et al., 2015). Similar high-resolution projections might be necessary to better assess exposure in other key coastal areas, including the Gulf Coast.

Second, as noted earlier, the FEMA flood maps applied in this analysis are incomplete, with some coastal areas lacking digitized BFEs and others working with maps updated in previous decades. FEMA's FIRMs are also limited in that their design is mainly for regulatory assessments versus risk analysis, and only include information at two exceedance probability intervals (100-year and 500-year). A systematic reassessment of coastal storm flood risk nationwide using updated methods would substantially improve upon the information available at present, but at present the only possible improvement would be a site-by-site, state-by-state review to identify the best available and most recent coastal flood risk estimates produced by local or state governments or through the academic and consulting community.

Finally, this assessment uses a simple "bathtub" assumption that applies RSLR rates linearly to FEMA BFE estimates. In practice, SLR and subsidence can actually produce storm surge and wave effects that increase non-linearly due to changes in surface friction and energy dissipation as coastal marshes and other natural barriers are submerged. As a result, the estimates here likely underestimate the potential storm surge flood risk with RSLR included for some coastal areas considered.

### 2.1.3. References

Burkett, Virginia, and Margaret Davidson, eds., *Coastal Impacts, Adaptation and Vulnerability: A Technical Input to the 2012 National Climate Assessment*, Cooperative Report to the 2013 National Climate Assessment, National Climate Assessment Regional Technical Input Report Series, Washington, D.C.: Island Press, 2012. As of March 24, 2015:
http://www.cakex.org/sites/default/files/documents/Coastal-NCA-1.13-web.form__0.pdf

Church, John A., and Neil J. White, "Sea-Level Rise from the Late 19th to the Early 21st Century," *Surveys in Geophysics*, Vol. 32, No. 4–5, September 1, 2011, pp. 585–602.

Crowell, Mark, Kevin Coulton, Cheryl Johnson, Jonathan Westcott, Doug Bellomo, Scott Edelman, and Emily Hirsch, "An Estimate of the U.S. Population Living in 100-Year Coastal Flood Hazard Areas," *Journal of Coastal Research*, 2010, pp. 201–211.

Federal Emergency Management Agency, "Hazus®-Mh 2.1 Technical and User's Manuals," 2012. As of March 24, 2015:
http://www.fema.gov/media-library/assets/documents/24609

———, "Flood Zones," web page, last updated April 27, 2016. As of June 6, 2016: https://www.fema.gov/flood-zones

FEMA—*See* Federal Emergency Management Agency

Horton, Randy, Celine Herweijer, Cynthia Rosenzweig, Jiping P. Liu, Vivien Gornitz, and Alex C. Ruane, "Sea Level Rise Projections for Current Generation Cgcms Based on the Semi-Empirical Method," *Geophysical Research Letters*, Vol. 35, No. 2, January 26, 2008.

Horton, Radley, Christopher Little, Vivien Gornitz, Daniel Bader, and Michael Oppenheimer, "New York City Panel on Climate Change 2015 Report Chapter 2: Sea Level Rise and Coastal Storms," *Annals of the New York Academy of Sciences*, Vol. 1336, No. 1, 2015, pp. 36–44.

Karl, Thomas, Gerald A. Meehl, Christopher D. Miller, Susan J. Hassol, Anne M. Waple, and William L. Murray, "Weather and Climate Extremes in a Changing Climate—Regions of Focus: North America, Hawaii, Caribbean, and U.S. Pacific Islands, Synthesis and Assessment Product 3.3," U.S. Climate Change Science Program, June 2008.

Kriebel, David L., and Joseph D. Geiman, "A Coastal Flood Stage to Define Existing and Future Sea-Level Hazards," *Journal of Coastal Research*, Vol. 30, No. 5, 2013, pp. 1017–1024.

Leifert, Harvey, "Sea Level Rise Added $2 Billion to Sandy's Toll in New York City," *EOS*, Vol. 96, March 16, 2015.

Miller, Kenneth G., Robert E. Kopp, Benjamin P. Horton, James V. Browning, and Andrew C. Kemp, "A Geological Perspective on Sea-Level Rise and Its Impacts Along the U.S. Mid-Atlantic Coast," *Earth's Future*, 2013.

National Oceanic and Atmospheric Administration Office for Coastal Management: Digital Coast, "Sea Level Rise Viewer," January 31, 2015. As of March 24, 2015: https://coast.noaa.gov/digitalcoast/tools/slr.html

National Oceanic and Atmospheric Administration /National Ocean Service, "Regional Mean Sea Level Trends," 2013a. As of March 23, 2015: http://tidesandcurrents.noaa.gov/sltrends/slrmap.htm

———, "Sea Level Trends," October 15, 2013b. As of March 24, 2015: http://tidesandcurrents.noaa.gov/sltrends/sltrends.html

NOAA—*See* National Oceanic and Atmospheric Administration.

Osborn, Timothy J., and Mike Hulme, "Development of a Relationship Between Station and Grid-Box Rainday Frequencies for Climate Model Evaluation," *Journal of Climate*, Vol. 10, No. 8, August 1997, pp. 1885–1908.

Pachauri, Rajendra K., and Andy Reisinger, eds., *Climate Change 2007: Synthesis Report: Contribution of Working Groups I, II and III to the Fourth Assessment Report of the Intergovernmental Panel on Climate Change*, Geneva, Switzerland: International Panel on Climate Change, 2007. As of March 24, 2015: http://www.ipcc.ch/publications_and_data/publications_ipcc_fourth_assessment_report_synthesis_report.htm

Parris, Adam, Peter Bromirski, Virginia Burkett, Dan Cayan, Mary Culver, John Hall, Radley Horton, Kevin Knuuti, Richard Moss, Jayantha Obeysekera, Abby Sallenger, and Jeremy Weiss, "Global Sea Level Rise Scenarios for the U.S. National Climate Assessment," National Oceanic and Atmospheric Administration Technical Memo OAR CPO-1, Climate Program Office, December 6, 2012. As of March 24, 2015: http://scenarios.globalchange.gov/sites/default/files/NOAA_SLR_r3_0.pdf

Spanger-Siegfried, Erika, Melanie Fitzpatrick, and Kristina Dahl, "Encroaching Tides: How Sea Level Rise and Tidal Flood Threaten U.S. and Gulf Coast Communities over the Next 30 Years," Union of Concerned Scientists, October 2014. As of March 24, 2015: http://www.ucsusa.org/sites/default/files/attach/2014/10/encroaching-tides-full-report.pdf

U.S. Geological Survey, "The National Map, 3D Elevation Program Products and Services," web page, 2014. As of June 3, 2016: http://nationalmap.gov/3dep_prodserv.html

Walsh, John, Donald Wuebbles, Katharine Hayhoe, James Kossin, Kenneth Kunkel, Graeme Stephens, Peter Thorne, Russell Vose, Michael Wehner, Josh Willis, David Anderson, Scott Doney, Richard Feely, Paula Hennon, Viatcheslav Kharin, Thomas Knutson, Felix Landerer, Tim Lenton, John Kennedy, and Richard Somerville, "Ch. 2: Our Changing Climate," in Jerry M. Melillo, Terese (T. C.) Richmond, and Gary W. Yohe, eds., *Climate Change Impacts in the United States: The Third National Climate Assessment: U.S. Global Change Research Program*, 2014, pp. 19–67. As of March 24, 2015:
http://nca2014.globalchange.gov/report/our-changing-climate/introduction

Woodruff, Jonathan D., Jennifer L. Irish, and Suzana J. Camargo, "Coastal Flooding by Tropical Cyclones and Sea-Level Rise," *Nature*, Vol. 504, No. 7478, 2013, pp. 44–52.

## 2.2. Extreme Temperature

Temperature extremes often are based either on the probability of occurrence or on a threshold value. In general, extreme temperature is defined as the number, percentage, or fraction of days with a maximum (or minimum) temperature, below (or above) the 1st, 5th, or 10th percentile or the 90th, 95th, or 99th percentile, for a given timespan (days, month, season, annual) with respect to a reference time frame. For the purposes of this analysis, we are concerned only with extreme heat. Specifically, we estimate from climate data the highest temperature a location is expected to experience every *T* years (*T* is the return period). We chose this metric to conduct our analysis for two reasons. First, this maximum temperature seems most likely to impact infrastructure, while measures of temperature over time and measures of nighttime low temperatures are geared toward assessing human health impacts.[4] Second, this measure is typically established using monthly temperature projections, which are much less computationally intensive to analyze than daily projections.

NOAA classifies U.S. temperature rankings as near normal (within the middle third of all periods on record), above normal (within top third), below normal (within bottom third), much-above (within the top tenth), or much-below (within the bottom tenth). Percentile rankings allow accounting for regional differences.

The impacts of increases in extreme temperature for infrastructure systems are primarily understood anecdotally. Rübbelke and Vögele (2011), for example, noted that a summer 2009 heat wave in France coupled with a drought resulted in cooling water shortages that put one-third of France's nuclear power plants out of service. The authors modeled how decreased water supply and warmer cooling water could limit nuclear power plant capacity to cool during peak summer periods. Smoyer-Tomic, Kuhn, and Hudson (2003) conducted a literature review to outline a list of the impacts of heat waves. Of relevance to this study are the potential impacts of heat waves on infrastructure assets, including

- increased damage to roads, railroad tracks, and bridges
- decreased efficiency in power transmission
- decreased power from hydropower and dams.

Smoyer-Tomic, Kuhn, and Hudson (2003) noted that most of the impacts of heat are relative to typical temperatures because people, plants, and animals adapt to changes in climate. On the other hand, impacts to infrastructure are absolute—they depend on the design of the infrastructure. Adaptation is probably most feasible with short-lived equipment needing replacement often (e.g., higher-capacity air conditioners that result in higher energy costs—and potentially higher risk if power grids cannot keep up with increasing temperatures). Such adaptations are more difficult with relatively long-lived infrastructure, but modifications and retrofits might be possible at some cost. Therefore, the adaptability of infrastructure over time is a major uncertainty. This study, with its focus on identifying regions and infrastructures to target for investments that can fund these infrastructure adaptations, will assume that infrastructure does not adapt from its current state.

---

[4] Long-lasting heat is primarily a worry for health, but it could also impact infrastructure that is slow to absorb heat. Given the lack of research on the infrastructure impacts of heat, we have elected to use a simple metric.

Of all hazards, the IPCC has found the most support for climate change increasing the likelihood and severity of extreme temperature. The IPCC's Special Report on Managing Risks of Extreme Events (SREX) states that since 1950 it is very likely that temperatures have increased on a global scale and with medium confidence finds that the length and number of warm spells have increased globally (Seneviratne et al., 2012, p. 111). The SREX finds that it is "virtually certain" that "the frequency and magnitude of warm daily temperature extremes" will occur in the future and very likely that the frequency, length, and intensity of heat waves will increase (Seneviratne et al., 2012, p. 112). For example, the once-in-20-year high temperature is likely to become a once-in-two-year high temperature across most areas (Seneviratne et al., 2012, p. 112).

But the various and complex mechanisms underlying increases in extreme temperatures make projections uncertain. Mechanisms leading to extreme temperature include atmospheric blocking, land-atmosphere feedbacks, soil moisture memory, snow feedbacks, and aerosols (Murray and Ebi, 2012). Drought conditions (and drier soil, which is a contributing factor to drought) can exacerbate temperature extremes (Andersen et al., 2005). Significantly, soil moisture and surface heating factors lead to changing mean temperature not always scaling with changes in temperature extremes (Haarsma, 2009; Murray and Ebi, 2012).

### 2.2.1. Data Sources

We use CMIP5 projections in our analysis of extreme temperature. One potential disadvantage of the CMIP5 projections is that they do not include measures of humidity. While humidity is a key component of apparent temperatures and hence a driver of mortality, in the context of this analysis infrastructure humidity is of less importance since most infrastructures are directly impacted by temperature only. Humidity can have important effects (for example, demand for electric power will increase with humidity and evaporative cooling of power plants will be less effective with higher levels of humidity), but high temperatures will primarily drive damage to infrastructure.

### 2.2.2. Analysis Methods

We use extreme value analysis software called extRemes for R to estimate extreme value models (The Weather and Climate Impact Assessment Science Program, 2014). We use the block maximum approach to estimate the model. This approach looks at each year (the block) and selects the highest temperature (the maximum) to estimate the extreme value models.[5] We estimate the default generalized extreme value (GEV) distribution, which is theoretically justified to fit the block maxima of data (see Gilleland and Katz, 2014, for an in-depth explanation).

Of the two methods generally used to estimate extreme value models, namely maximum likelihood and likelihood-moments estimation, we use the latter, less computationally intensive option because we are calculating tens of thousands of these models.

We project the potential vulnerabilities from climate-induced increases in extreme heat by applying a range of temperature thresholds to climate projections and identifying when

[5] An alternative is the Peak-over-Threshold Approach, which uses any peaks over a certain threshold. The advantage of this method is that it produces smaller confidence intervals. One disadvantage is that estimation of extreme-value models requires independent events, but consecutive days of high temperature are really a single event; thus preparing the data for analysis can be difficult. The other disadvantage is that this estimation would require daily data, which would increase computational demands tremendously.

and where temperatures exceed those thresholds. For example, we might be interested in flagging infrastructure where the 50-year exceedance temperature—that is, the temperature that is likely to be exceeded on average only once in 50 years—was 130 degrees or higher if a 2 percent AEP of reaching temperatures higher than 130 degrees was deemed an unacceptable risk. Our analysis examines where extreme temperatures are expected to exceed two thresholds—120 degrees and 130 degrees. The data used could be adapted to analyze other temperature thresholds.

Table 2.6 outlines the procedure we use to estimate climate-induced changes to extreme heat.

Figures 2.3 and 2.4 conclude this section of the chapter on extreme temperature by showing example calculations of the 100-year extreme temperatures forecast for 2100, and the

**Table 2.6**
**Estimation of Climate-Induced Changes to Extreme Heat**

| Analysis Factor | Description |
|---|---|
| Years | 2005–2015 (to represent current conditions, i.e., 2010)<br>2035–2045 (to represent conditions in 2040)<br>2060–2070 (to represent conditions in 2065)<br>2090–2099 (to represent the end of the century, i.e., 2100) |
| Data | CMIP5 Climate Data, 1/8 degree<br>• Monthly maximum temperature |
| Scenarios | • RCP 4.5<br>• RCP 8.5 |
| Return periods | 2/5/10/20/50/100 years |
| Models | Projections performed using ensemble estimates of all models combined together. Ensemble estimates will be used because climate models are relatively consistent in their projections of temperature, whereas projections of precipitation are highly varied. Combining all the data into an ensemble is advantageous because it increases the sample size from which to estimate extreme value models. If the data were not combined, there would be relatively little data on which to estimate models since there are only ten years of data, and most places in the United States will experience their maximum temperatures only during the summer, hence months in other seasons should have little effect on extreme values.<br>• RCP 4.5: 21 models<br>• RCP 8.5: 21 models<br>Note: GCM data was not available for analysis of extreme temperature using the MIROC5_r2i1p1 model for any time periods or for the bcc.csm1.1_r1i1p1 model in the 2100 time period. |
| Thresholds | • 120 degrees<br>• 130 degrees |
| Procedure | For each combination of 1/8 degree cell and scenario, the following procedures were performed:<br>• Combine the maximum monthly temperature for the full ensemble of models. For example, combining data from the 22 models in RCP 4.5 creates a time series that is 220 years long.<br>• Estimate an extreme value model. Modeling was conducted with extreme value analysis software called extRemes (The Weather and Climate Impact Assessment Science Program, 2014).<br>• Using the estimated model, calculate the temperature for each of the exceedance periods<br>• Compare exceedance temperatures with vulnerability thresholds and locations of infrastructure. |

change in these extreme temperatures from the present day. Areas with a high-forecast temperature in Figure 2.3 will have the most temperature stress on infrastructure, but many of those areas are already hot. Areas with the biggest changes in Figure 2.4 are likely to have the most difficulty adapting to higher temperatures.

**Figure 2.3**
**Extreme Temperature for 100-Year Return, RCP 8.5, 2100**

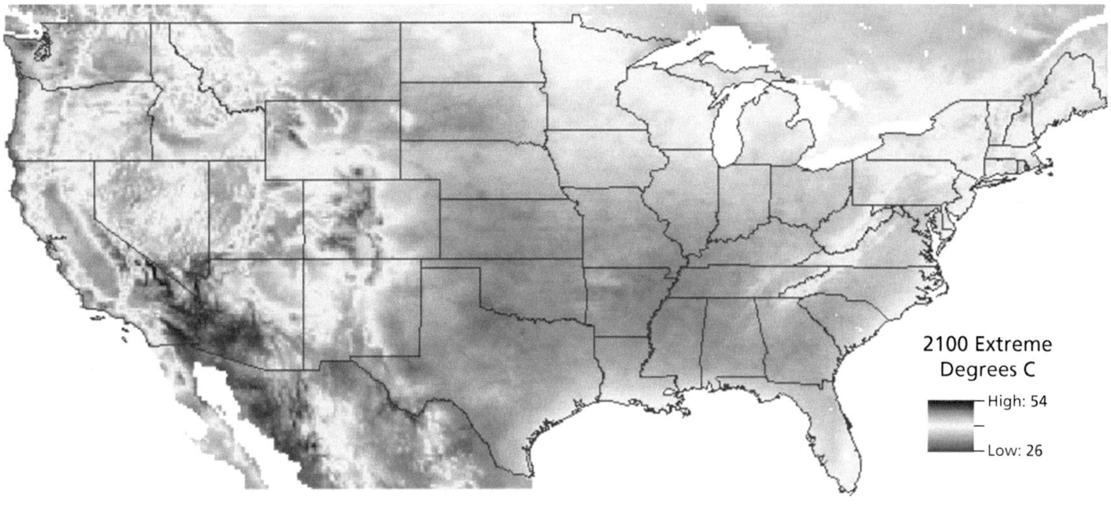


SOURCE: RAND analysis of ensemble of all CMIP5 models.
RAND *RR1453/1-2.3*

**Figure 2.4**
**Change from 2010 to 2100 of Extreme Temperature for 100-Year Return, RCP 8.5**

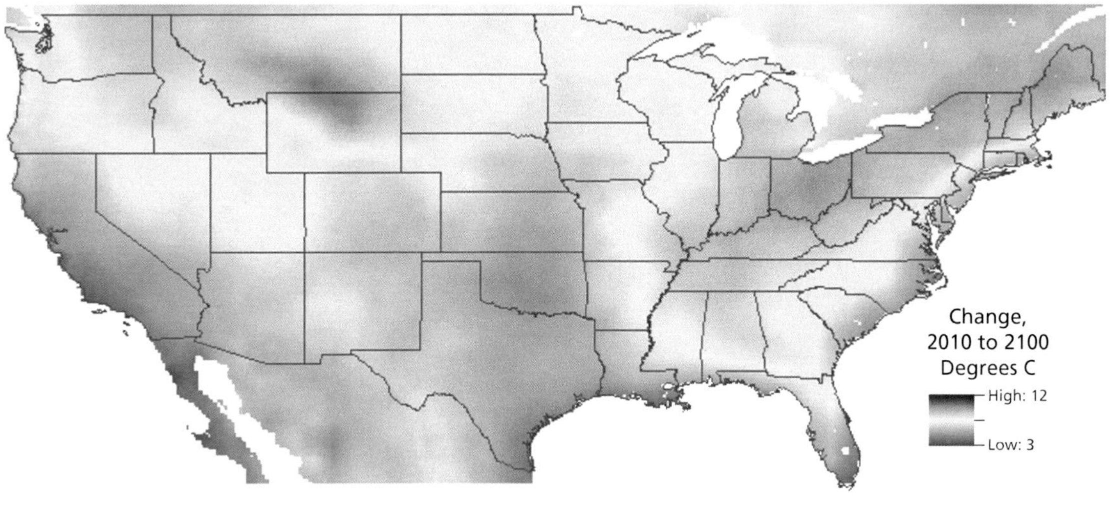


SOURCE: RAND analysis of ensemble of all CMIP5 models.
RAND *RR1453/1-2.4*

### 2.2.3. References

Andersen, Ole B., Sonia I. Seneviratne, Jacques Hinderer, and Pedro Viterbo, "GRACE-Derived Terrestrial Water Storage Depletion Associated with the 2003 European Heat Wave," *Geophysical Research Letters*, Vol. 32, No. 18, 2005.

Gilleland, Eric, and Richard W. Katz, "extRemes 2.0: An Extreme Value Analysis Package in R," submitted to the *Journal of Statistical Software*, June 3, 2014. As of April 12, 2016: http://www.ral.ucar.edu/~ericg/extRemes/extRemes2.pdf

Haarsma, Reindart J., Frank Selten, Bart vd Hurk, Wilco Hazeleger, and Xueli Wang, "Drier Mediterranean Soils Due to Greenhouse Warming Bring Easterly Winds Over Summertime Central Europe," *Geophysical Research Letters*, Vol. 36, No. 4, 2009.

Murray, Virginia, and Kristie L. Ebi, "IPCC Special Report on Managing the Risks of Extreme Events and Disasters to Advance Climate Change Adaptation (SREX)," *Journal of Epidemiology and Community Health*, Vol. 66, No. 9, September 2012, pp. 759–760.

Rübbelke, Dirk, and Stefan Vögele, "Impacts of Climate Change on European Critical Infrastructures: The Case of the Power Sector," *Environmental Science & Policy*, Vol. 14, 2011, pp. 53–63.

Seneviratne, Sonia, Neville Nicholls, David Easterling, Clare M. Goodess, Shinjiro Kanae, James Kossin, Yali Luo, Jose Marengo, Kathleen McInnes, Mohammad Rahimi, Markus Reichstein, Asgeir Sorteberg, Carolina Vera, and Xuebin Zhang, "Changes in Climate Extremes and Their Impacts on the Natural Physical Environment," in Intergovernmental Panel on Climate Change, eds., *Managing the Risks of Extreme Events and Disasters to Advance Climate Change Adaptation*, New York: Cambridge University Press, 2012, pp. 109–230.

Smoyer-Tomic, Karen E., Robyn Kuhn, and Alana Hudson, "Heat Wave Hazards: An Overview of Heat Wave Impacts in Canada," *Natural Hazards*, Vol. 28, 2003, pp. 463–485.

The Weather and Climate Impact Assessment Science Program, "Extreme Value Analysis Software," web page, July 2014. As of April 12, 2016:
http://www.assessment.ucar.edu/toolkit/

## 2.3. Meteorological Drought

Drought is usually defined relative to a region's historical climate using one of several metrics, some of which are outlined in the analysis section of this report. Meteorological drought refers to conditions when dry weather patterns dominate an area. This type of drought is in contrast to hydrological drought (i.e., when low water supply becomes evident), agricultural drought (i.e., when crops are affected), or socioeconomic drought (i.e., when the supply and demand of commodities are affected).

Because human activities across geographies have evolved under pressure from climate conditions, infrastructure has been adapted to meet local climate conditions. The specific conditions for drought (e.g., rainfall and temperature) will vary across geography because of differences in climate. For example, a wet year in a typically arid desert could be a year of extreme drought for a region along the Gulf Coast.

Drought can have an especially large impact on agriculture. For example, Backus, Lowry, and Warren (2013) estimated the potential economic impacts from climate–change-induced drought through 2050. In these models, decreases in agricultural productivity from water deficiencies resulted in some of the largest economic losses from climate-induced drought in the U.S. economy due to higher transportation costs to agriculture-demanding industries (see Warren et al., 2010). But beyond the agricultural sector, drought could also have notable impacts for electric power generation and transmission systems. In 2007, drought in the southeastern U.S. caused nuclear and coal-fired plants in the Tennessee Valley Authority system to curtail operations. In 2006, nuclear plants in Illinois and Minnesota were affected by drought affecting water levels in the Mississippi River (Argonne National Laboratory, 2012).

Because drought is so closely linked to climate, changes in drought brought about by climate change can be more worrying than typical climatic variability because infrastructure has evolved under the pressures of the typical variability. For example, the Drought Severity Index in Aqueduct Global Maps 2.0 (Gassert et al., 2013) considers a drought to be any time soil moisture dips below the 20th percentile of observed soil moisture observations from 1901 to 2008. By this definition of drought, all regions have historically been in a drought 20 percent of the time in that period. These droughts vary by severity (the length of time soil moisture is below the 20th percentile before it increases, and the relative dryness of the drought relative to average conditions), but they have occurred everywhere with the same frequency. As climate changes, what used to be the 20th percentile for soil moisture can change. Given this historical definition of a drought, some regions can be in drought more frequently than others as climate changes. Furthermore, the relative severity of those droughts might change.

### 2.3.1. Data Sources

The primary source of data for the future climate-influenced changes to dryness is CMIP5 projections (Lawrence Livermore National Laboratories, 2013). These data contain various models of future climate that simulate future daily temperature and precipitation across various IPCC climate scenarios.

Each climate projection is a result of combining a GCM with a forcing emissions scenario. CMIP5 projections used the latest version of the GCMs and an updated set of emissions scenarios. In this analysis, 22 different GCMs covering the period from 2006 to 2100 and two emissions scenarios (RCP 4.5 and RCP 8.5) are considered. The RCPs reflect advancements in integrated assessment modeling to characterize future developments in global greenhouse gas emissions. The data considered in the study for these projections cover the time period from 2006 to 2100.

### 2.3.2. Analysis Methods

This study's projections of soil dryness use CMIP5 data. CMIP5 projections consider soil moisture and net surface water changes (defined as the precipitation minus evapotranspiration [P-ET]) (Wuebbles et al., 2013). Critical to modeling drought risk, the CMIP5 simulations of the North American monsoon are an improvement over previous versions that failed to adequately simulate all circulation patterns (Wuebbles et al., 2013). Despite large inter-model variations, CMIP5 comparisons are consistent with offline hydrology models (Wuebbles et al., 2013).

We use the KBDI to measure projected changes in dryness. Calculation of projections for this index requires measures of daily maximum temperature and daily precipitation—measures that can be calculated with data available from CMIP5 for all of CONUS. KBDI estimates for the water content of soil/duff are based on daily temperature and precipitation observations over time. Originally developed to estimate the likelihood of forest fires in the southeastern United States, KBDI is more generally applicable to measuring dryness. KBDI values range from 0 (completely saturated soil) to 800 (completely dry soil), scaled to assume that soil can hold from 0 to 8 inches of water (Keetch and Byram, 1968).

A number of other measures are used in the literature to measure lack of precipitation and soil dryness. These measures can be broken into those that only require data about temperature and precipitation (of which KBDI is one) and those that require other data, such as evapotranspiration, soil moisture, and runoff. Tables 2.7 and 2.8 provide a list and brief descriptions of some of these other methods.

**Table 2.7**
**Drought Indexes: Measures of Dryness Using Temperature or Precipitation**

| | |
|---|---|
| Standard Precipitation Index | Based on standardized precipitation data to a normal distribution *N(0,1)* and calculated on the number of selected time periods (1–48 months):<br>• –3 (extreme droughts)<br>• –1.5 to –2 (severe droughts)<br>• –1 to –1.5 (moderate droughts)<br>• –0.5 to –1 (mild droughts)<br>• 0 to +2 (mild or severely wet)<br>• +3 (extremely wet)<br>(Edwards, 1997; Lloyd-Hughes and Saunders, 2002) |
| Consecutive dry days (CDD) | Based on the maximum consecutive number of days of no rain, below a given threshold (e.g., <1 mm per day) within a considered period (e.g., 1 year) (Tebaldi et al., 2006) |
| Crop Moisture Index (CMI) | Based on mean temperature and total precipitation, from the existing and previous week, to measure short-term conditions:<br>• –3 (severely dry)<br>• +3 (excessively wet)<br>(Palmer, 1968) |
| No-rain episodes | Duration of no-rain periods (e.g., >20, 30, or 60 days) (Groisman and Knight, 2008) |

**Table 2.8**
**Drought Indexes: Measures of Dryness Using Temperature, Precipitation, Evapotranspiration, Soil Moisture, or Runoff**

| | |
|---|---|
| Palmer Drought Severity Index (PDSI) | Based on precipitation and potential evapotranspiration, measures the difference of moisture balance from normal conditions using a simple water balance model:<br>• −4 (extreme drought)<br>• +4 (extreme wet)<br>(Palmer, 1965) |
| Precipitation Potential Evaporation Anomaly | Based on precipitation and potential evapotranspiration, looks at the cumulative difference between precipitation and potential evapotranspiration (Burke, Perry, and Brown, 2010) |
| Standardized Precipitation-Evapotranspiration Index (SPEI) | Based on cumulated anomalies of precipitation and potential evapotranspiration (Vicente-Serrano, Begueria, and Lopez-Moreno, 2010) |
| Simulated Soil Moisture Anomalies | Integrated effects of precipitation forcing, simulated actual evapotranspiration, and simulated soil moisture persistence (Murray and Ebi, 2012) |
| Soil Moisture and Runoff Index | Based on normalized model-derived soil moisture and runoff:<br>• 0 (dry)<br>• 1 (wet) |

We use frequency statistics instead of exceedance metrics because exceedance metrics are not compatible with the way that KBDI is scaled. According to the U.S. Forest Service's Wildland Fire Assessment System (2014), KBDI has four categories: typical early spring (KBDI of 0 to 200), typical late spring (KBDI of 200 to 400), typical late summer (KBDI of 400 to 600), and severe drought (KBDI of 600 to 800). In other words, most of the KBDI scale typically will be reached during a year somewhere in the United States because of seasonal variation. KBDI is not a distribution with a long tail like many of the other hazards—it has a fixed maximum that is often reached. Therefore, we calculate frequency statistics, which will be impacted by changes in the length of dry spells (e.g., longer dry spells will increase the frequency of high values of KBDI) and the severity of dry spells (e.g., more severe dry spells will increase KBDI). The 75th percentile will be driven by dryness during the dry season. The 95th percentile will be driven by the most extreme dryness events.[6] The procedure to model climate-induced changes in dryness is outlined in Table 2.9.

Figure 2.5 shows the calculations for the 75th percentile KDBI value for one model for the RCP 8.5 scenario for 2090 to 2099, i.e., the scenario and time period expected to be most impacted by climate change. The 75th percentile is roughly the average minimum KBDI during the dry season. Figure 2.6 clarifies the 400 and 600 cutoff points; on average, areas in yellow are in a severe drought for at least three months of the year, while areas in red are in an extreme drought for at least three months of the year.

[6] We also calculated the average and 50th percentile, but we do not plan to use them as thresholds.

**Table 2.9**
**Estimation of Climate-Induced Changes to Soil Dryness**

| Analysis Factor | Description |
|---|---|
| Years | 2006–2020 (to represent current conditions)<br>2035–2045 (to represent conditions 25 years in the future, i.e., 2040)<br>2060–2070 (to represent conditions 50 years in the future, i.e., 2065)<br>2090–2100 (to represent the end of the century, i.e., 2100) |
| Data | CMIP5 climate data, 1/8 degree<br>• Daily maximum temperature<br>• Daily precipitation |
| Scenarios | • RCP 4.5<br>• RCP 8.5 |
| Models | Projections will be performed for each model separately because of the variance in precipitation projections across models:<br>• RCP 4.5: 22 models<br>• RCP 8.5: 22 models |
| Return periods | • 75th percentile (which corresponds to the minimum KBDI for the driest three months of the year)<br>• 95th percentile (which is a measure of the most-extreme drought events) |
| Thresholds | • Severe drought (KBDI of 600 to 800)<br>• Dry or severe drought (KBDI of 400 to 800) |
| Model aggregation | • 25th percentile of KBDIs across model runs<br>• Median (50th percentile) of KBDIs across model runs<br>• 75th percentile of KBDIs across model runs<br>• 95th percentile of KBDIs across model runs<br>• Mean KBDIs across model runs<br>• Maximum KBDIs across model runs |
| Procedure | For each combination of 1/8 degree cell and scenario/model, the following procedures were performed:<br>1. Calculate the daily KBDI<br>• Assume a starting KBDI of zero[a]<br>• Use the "Fire Danger Index Functions in R" software to calculate KDBI (Williamson, 2010).<br>• Use the "index_KBDI" function, which requires temperature, precipitation, and the mean annual precipitation (calculated from baseline data)[b]<br>2. Calculate frequency statistics<br>3. Compare KBDI statistics with thresholds and locations of infrastructure |

[a] A KBDI of zero assumes that the soil is saturated at the beginning of the period. Ideally, KBDI would be calculated across all years—including historical years—for each model to assure that KBDI always reflects historical conditions. However, we assume a starting KBDI of zero to compensate for high computational requirements of calculating KBDI across all years. We believe this assumption has minimal impact on calculations of the 75th and 95th percentile because the soil in most U.S. locations is relatively saturated on January 1 and generally becomes more saturated over the winter because there is little evaporation.

[b] When implementing this modeling, we discovered that the code was extremely slow, so it could not be used to do a large number of calculations as would be needed in our modeling. Therefore, we rewrote a table lookup function to make it faster and we changed how some of the calculations were using memory, but we did not make any material changes in the calculations.

**Figure 2.5**
**Minimum KBDI for Driest Three Months of the Year (75th percentile) Using the CanESM2 r2i1p1 Model, RCP 8.5 Scenario, for 2090–2099**

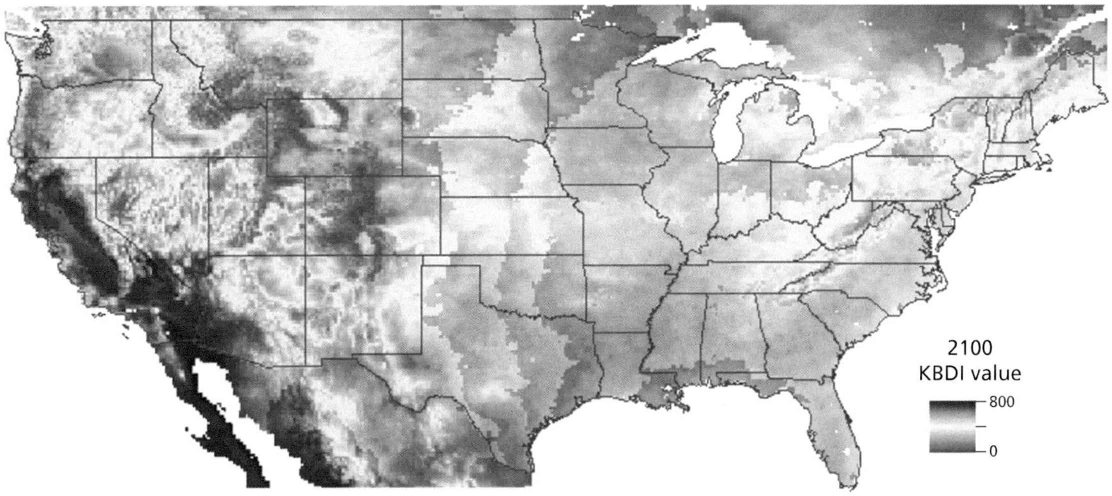


RAND *RR1453/1-2.5*

**Figure 2.6**
**Severe and Extreme Drought Cutoffs Using the CanESM2 r2i1p1 Model, RCP 8.5 Scenario, for 2090–2099**

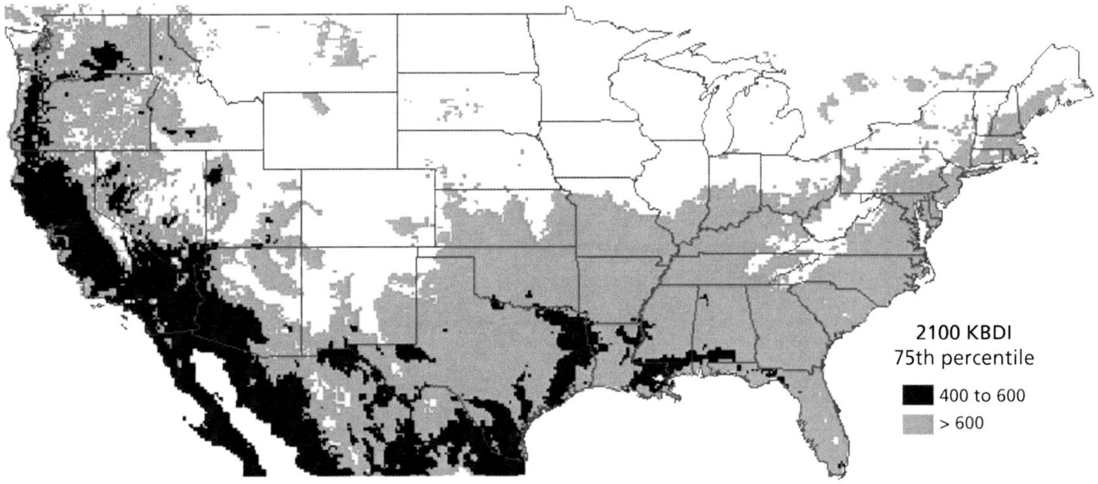


NOTE: Areas in orange are in a typical late-summer drought, and areas in red are in a severe drought, for at least three months of the year. There is stronger agreement between models on long-term drought and large spatial resolution, while there is more variability of shorter-term drought on a regional scale (Blenkinsop and Fowler, 2007). Two important uncertainties concerning future drought trends include ocean circulation and land-atmosphere feedback interactions (e.g., drought impacts on vegetation physiology) (Murray and Ebi, 2012). Lack of observations and full representation of soil moisture-evapotranspiration factor into climate models and to existing uncertainties (Seneviratne et al., 2010).

RAND *RR1453/1-2.6*

### 2.3.3. References

Argonne National Laboratory, *Impacts of Long-Term Drought on Power Systems in the U.S. Southwest*, prepared for U.S. Department of Energy, Office of Electric Delivery and Energy Reliability, 2012. As of April 12, 2016: http://energy.gov/sites/prod/files/Impacts%20of%20Long-term%20Drought%20on%20Power%20Systems%20in%20the%20US%20Southwest%20–%20July%202012.pdf

Backus, George A., Thomas S. Lowry, and Drake E. Warren, "The Near-Term Risk of Climate Uncertainty Among the U.S. States," *Climatic Change*, Vol. 116, No. 3–4, 2013, pp. 495–522.

Blenkinsop, Stephen, and Hayley J.Fowler, "Changes in European Drought Characteristics Projected by the PRUDENCE Regional Climate Models," *International Journal of Climatology*, Vol. 27, No. 12, 2007, pp. 1595–1610.

Burke, Eleanor J., Richard H. J. Perry, and Simon Brown, "An Extreme Value Analysis of UK Drought and Projections of Change in the Future," *Journal of Hydrology*, Vol. 388, No. 1–2, 2010, pp. 131–143. As of April 12, 2016:
https://www.researchgate.net/publication/222919988_An_extreme_value_analysis_of_UK_drought_and_projections_of_change_in_the_future

Edwards, Daniel C., "Characteristics of 20th Century Drought in the United States at Multiple Time Scales," master's thesis, Defense Technical Information Center, May 1997. As of April 12, 2016:
http://www.dtic.mil/dtic/tr/fulltext/u2/a325595.pdf

Gassert, Francis, Matt Luck, Matt Landis, Paul Reig, and Tien Shiao, "Aqueduct Global Maps 2.0," working paper, Washington, D.C.: World Resources Institute, January 2013. As of April 12, 2016:
http://wri.org/publication/aqueduct-global-maps-20

Groisman, Pavel Ya, and Richard Knight, "Prolonged Dry Episodes over the Conterminous United States: New Tendencies Emerging During the Last 40 Years," *Journal of Climate*, Vol. 21, No. 9, 2008.

Keetch, John J., and George Byram, *A Drought Index for Forest Fire Control*, research paper SE-38, Asheville, N.C.: U.S. Department of Agriculture, Forest Service, Southeastern Forest Experiment Station, 1968. As of April 12, 2016:
http://www.srs.fs.usda.gov/pubs/viewpub.jsp?index=40

Lawrence Livermore National Laboratories, "Downscaled CMIP3 and CMIP5 Climate and Hydrology Projections: About," web page, May 7, 2013, last updated July 25, 2014. As of April 12, 2016:
http://gdo-dcp.ucllnl.org/downscaled_cmip_projections/dcpInterface.html#About

Lloyd-Hughes, Benjamin, and Mark Saunders, "A Drought Climatology for Europe," *International Journal of Climatology*, Vol. 22, No. 13, 2002, pp. 1571–1592.

Murray, Virginia, and Kristie L. Ebi, "IPCC Special Report on Managing the Risks of Extreme Events and Disasters to Advance Climate Change Adaptation (SREX)," *Journal of Epidemiology and Community Health*, Vol. 66, No. 9, September 2012, pp. 759–760.

Palmer, Wayne C., "Meteorological Drought," research paper No. 45, Washington, D.C., U.S. Department of Commerce, Weather Bureau, 1965.

———, "Keeping Track of Crop Moisture Conditions, Nationwide: The New Crop Moisture Index," *Weatherwise*, Vol. 21, No. 4, 1968, pp. 156–161.

Seneviratne, Sonia I., Thierry Corti, Edouard L. Davin, Martin Hirschi, Eric B. Jaeger, Irene Lehner, Boris Orlowsky, and Adriaan J. Teuling, "Investigating Soil Moisture-Climate Interactions in a Changing Climate: A Review," *Earth-Science Reviews*, Vol. 99, No. 3–4, 2010, pp. 125–161. As of April 12, 2016:
http://www.sciencedirect.com/science/article/pii/S0012825210000139

Tebaldi, Claudia, Katharine Hayhoe, Julie M. Arblaster, and Gerald A. Meehl, "Going to the Extremes," *Climatic Change*, Vol. 79, No. 3–4, 2006, pp. 185–211.

Vicente-Serrano, Sergio M., Santiago Begueria, and Juan I. Lopez-Moreno, "A Multiscalar Drought Index Sensitive to Global Warming: The Standardized Precipitation Evapotranspiration Index," *Journal of Climate*, Vol. 23, No. 7, 2010, pp. 1696–1718.

Warren, Drake E., Mark A. Ehlen, Verne W. Loose, and Vanessa N. Vargas, *Estimates of the Long-Term U.S. Economic Impacts of Global Climate Change-Induced Drought*, SAND Report SAND2010-0692, Sandia National Laboratories, 2010.

Wildland Fire Assessment System, "Keetch-Byram Drought Index," U.S. Forest Service, accessed August 27, 2014. As of April 12, 2016:
http://www.wfas.net/index.php/keetch-byram-index-moisture--drought-49

Williamson, Grant, "Fire Danger Index Functions in R," web page, updated August 16, 2010. As of April 12, 2016:
http://www.atriplex.info/index.php/Fire_Danger_Index_Functions_in_R

Wuebbles, Donald, Gerald Meehl, Katharine Hayhoe, Thomas R. Karl, Kenneth Kunkel, Benjamin Santer, Michael Wehner, Brian Colle, Erich M. Fischer, Rong Fu, Alex Goodman, Emily Janssen, Viatcheslav Kharin, Huikyo Lee, Wenhong Li, Lindsey N. Long, Seth C. Olsen, Zaitao Pan, Anji Seth, Justin Sheffield, and Liqiang Sun, "CMIP5 Climate Model Analyses: Climate Extremes in the United States," *Bulletin of the American Meteorological Society*, Vol. 95, 2013, pp. 571–583.

## 2.4. Wildfires

Wildfire impacts range from property damage, to physical and psychological harm. Economic costs include property damage, reduction in tourism, and timber losses (Morton et al., 2003). Forest fires impact mortality and increase the incidence of respiratory and cardiac illnesses (Gamble et al., 2008). Indirect health impacts can further occur from increased risk of landslides or soil erosion (McMichael et al., 2003). At the same time, there are often resources, such as insurance and private donations, made available to recover from this hazard (Banks et al., 2012).

Fire risk can be measured by length, frequency, and severity, as well as by a decrease in fire extinguishment and faster fire spread (Murray and Ebi, 2012). Fire risk can increase because of changes in climate extremes, including drought, low humidity, and high temperatures. For example, droughts can turn vegetation into wildfire fuel. Combined with human sources of ignition (i.e., deforestation), these variables simultaneously can change fire risk levels from low risk to medium risk and medium risk to high risk (Murray and Ebi, 2012). However, no one process directly describes an increase in fire occurrence.

The SREX cites studies that show drought and increased temperatures in North America are linked to increased risk of wildfires, and that wildfire activity has increased substantially since 1950 (Handmer et al., 2012, pp. 252, 259). For example, some research shows that higher temperatures lead to earlier snowmelt, which increases the length of the fire season. The SREX finds that people and infrastructure have been moving into more vulnerable areas. For example, the movement of urban areas into the bush exacerbated the impact of the 2009 Australian fires (Handmer et al., 2012, p. 239).

### 2.4.1. Data Sources

We use CMIP5 climate data to adjust the Wildland Fire Potential (now called Wildfire Hazard Potential) (USDA, 2012) based on projected changes in the KBDI index calculated for drought as described in the previous section.

### 2.4.2. Analysis Methods

Current, steady-state (i.e., not accounting for present year-to-year changes in drought conditions) wildfire risk for the United States is estimated by the Wildland Fire Potential from the U.S. Department of Agriculture, Forest Service, Fire Modeling Institute. The methodology for creating the Wildland Fire Potential map is described in Dillon, Menakis, and Fay (2013). To create the map, the researchers estimated the likelihood of wildfires and wildfire intensity using the Large Fire Simulator (FSim) (Finney et al., 2011). FSim simulates wildfires using a 10,000- to 50,000-year simulation driven by simulated weather. Many areas of the country have a relatively high annual probability of burning (up to 10 percent), but many of the fires would be small and relatively easy to control, thus not a major threat to infrastructure. The Wildland Fire Potential map calculates the risk of combining the likelihood of any wildland fire starting with the likelihood that long flame lengths would result in a crown fire that burns the entire forest (rather than only material on the ground or at low heights). To account for the consequences, the researchers weigh crown fires and surface fires based on fire intensity. Finally, the researchers adjust these consequences to account for the ease of controlling fires, which varies across different fuel types.

**Figure 2.7**
**Wildland Fire Potential Risk Map**

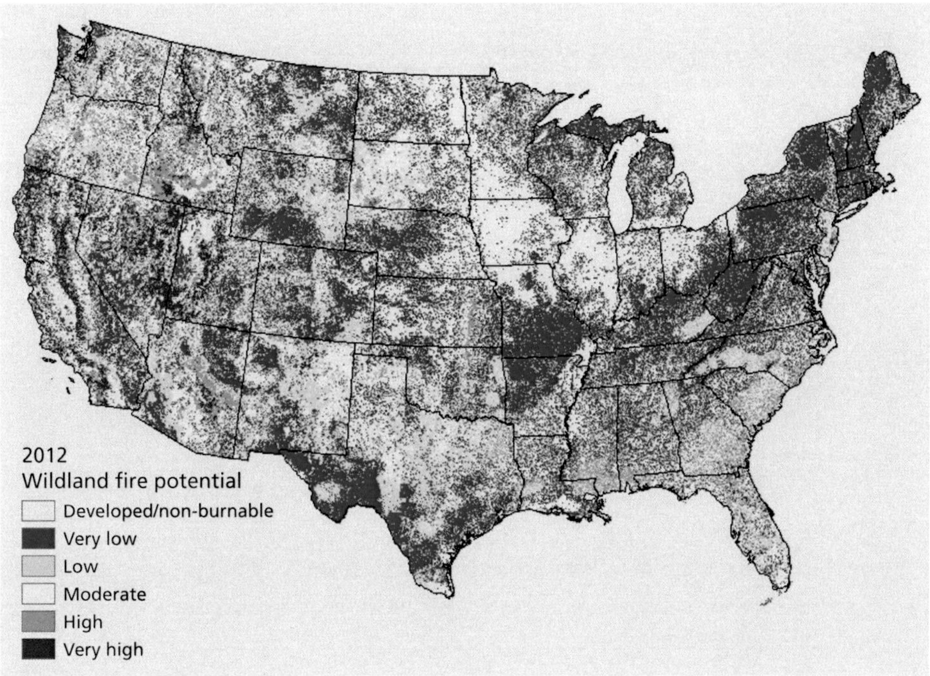


SOURCE: USDA, 2012.
RAND *RR1453/1-2.7*

The Wildland Fire Potential and FSim projections do not account for short-term variations in weather or fuel moisture, thus they are long-term projections based on historical weather observations measured over the past 10 years to 30 years by the National Fire Danger Rating System (NFDRS) Remote Automated Weather Stations. Therefore, the projections are long term—when moisture content and weather conditions are more favorable for fires, true fire risk will exceed the projections; and when moisture content and weather conditions are less favorable for fires, true fire risk will be below the projections.

Thompson et al. (2011) offer a similar, but more complex methodology for calculating wildfire risk. As in the Wildland Fire Potential projections, they use the FSim projections of likelihood and intensity. Rather than integrating these projections into a measurement of risk, such as the Wildland Fire Potential, they apply a set of weights that measure the damage that different flame lengths cause to various types of infrastructure to estimate the average annual loss from wildfires.[7]

Liu, Stanturf, and Goodrick (2010) forecast climate-induced changes in future Wildland Fire Potential indexes using the KBDI index. They use four IPCC scenarios (A1, A2, B1, and B2) and the results for four GCMs to forecast worldwide changes in KBDI indexes. Because

---

[7] Subject-matter experts determined the weights of eight different types of infrastructures or environmental amenities. Low-intensity fires were assumed to benefit some infrastructures and environmental amenities (e.g., ski areas), while having little effect on most infrastructure and environmental amenities (e.g., energy infrastructure), and having a near total loss for residential housing. High-intensity fires were assumed to be very destructive to most infrastructures and environmental amenities.

the GCM models are coarse, the projections of the indexes are coarse, too. The authors summarize the results for four regions of the United States (southwest, northwest, southeast, and northeast). Average projected increases in KBDI range from 60 (northeast, B1 scenario) to 240 (northwest, A1 scenario). Liu, Goodrick, and Stanturf (2013) builds on Liu Stanturf, and Goodrick (2010) by focusing on the United States and looking at a more-detailed set of regions. This follow-on study also looks at two weather factors—wind speed and relative humidity—that are not accounted for by the KBDI index. The study calculates KBDI separately for four seasons.

The Wildland Fire Potential (WFP) is an index that is calculated by integrating measures of severity (the occurrence of various intensities of flame and crown fires) with likelihood. Therefore, it accounts for both likelihood and consequence, so index cutoffs and thresholds are for overall wildfire risk.

Table 2.10 shows the cutoffs that were used to classify WFP values. WFP values are approximately distributed log-normally, and the cutoffs for the levels are approximately one log-unit above each other (i.e., a factor of 2.7 more than the previous threshold). Dillon et al. (2013) detail the procedure for setting the classification cutoffs.

To identify infrastructure at risk of wildland fires, we choose a threshold in the risk index (e.g., 401, which represents the cutoff for "moderate" risk) and identify all infrastructure that touches a cell in the Wildland Fire Potential map above that risk index.[8] This cutoff can vary by type of infrastructure; for example, a type of infrastructure that is more vulnerable to fire could have a lower cutoff.

The climate projections in this analysis estimate changes in wildfire risk by adjusting the Wildland Fire Potential (or the FSim likelihood) by a factor based on projections of the KBDI (as computed in the climate projections of meteorological and agricultural drought). KBDI can be interpreted as a measure of wildfire potential based upon temperature and precipitation over time that ranges from 0 (representing fully saturated soil) to 800 (representing total moisture deficiency).

A fundamental difficulty in applying these changes to fire potential to the present-day WFP index is that there is no direct link between KBDI and either the likelihood of a fire

**Table 2.10**
**Categorization Thresholds for Wildland Fire Potential**

| WFP Classification | Max WFP | log(Max WFP) |
|---|---|---|
| Nonburnable lands/water | N/A | N/A |
| Very Low | 51 | 3.9 |
| Low | 155 | 5.0 |
| Moderate | 401 | 6.0 |
| High | 1,935 | 7.6 |
| Very High (Max) | 98,080 | 11.5 |

NOTE: WFP = Wildland Fire Potential.

[8] Each cell, or pixel, in the Wildland Fire Potential map is approximately 270 meters by 270 meters.

occurring or the intensity of fires increasing. The research from Liu, Stanturf, and Goodrick (2010) and Liu, Goodrick, and Stanturf (2013) make clear that conditions are expected to become more favorable for wildfires with climate change. Furthermore, wildfire seasons are expected to lengthen, which would further increase the likelihood of wildfires. Given the limitations of these data, we use the simple assumption that the projected change in KBDI scales to increase the Wildland Fire Potential index (Keetch and George, 1968).[9]

Ideally, we would want to calculate the likelihood and intensity of wildfires as a function of weather inputs, such as temperature and precipitation and the changes in each. However, literature explaining these relationships does not exist. The KBDI index indicates the state of meteorological drought, which is one factor that is associated with increased wildfire risk. However, until research is available to connect weather factors directly to wildfire indexes, we are left to rely on observations that correlate the KBDI index to wildfire indexes.

Figure 2.8 shows the change in the 75th percentile KBDI from present day to about 2100 that is forecast by the CanESM2 r2i1p1 model for the RCP 8.5 scenario. (This is the same example used in the agricultural drought section.) The model forecasts increased drought in most of the country. These changes are most extreme in the northeast and northwest—as well as scattered other areas—where they exceed a one-level increase in fire risk (i.e., an increase in KBDI of 200, as used in Liu, Stanturf, and Goodrick, 2010). This particular model forecasts moderate decreases in KBDI for much of the interior western United States.

The procedure to model climate-induced changes in wildfire is outlined in Table 2.11.

**Figure 2.8**
**Projected Change in KBDI; CanESM2 r2i1p1 Model**

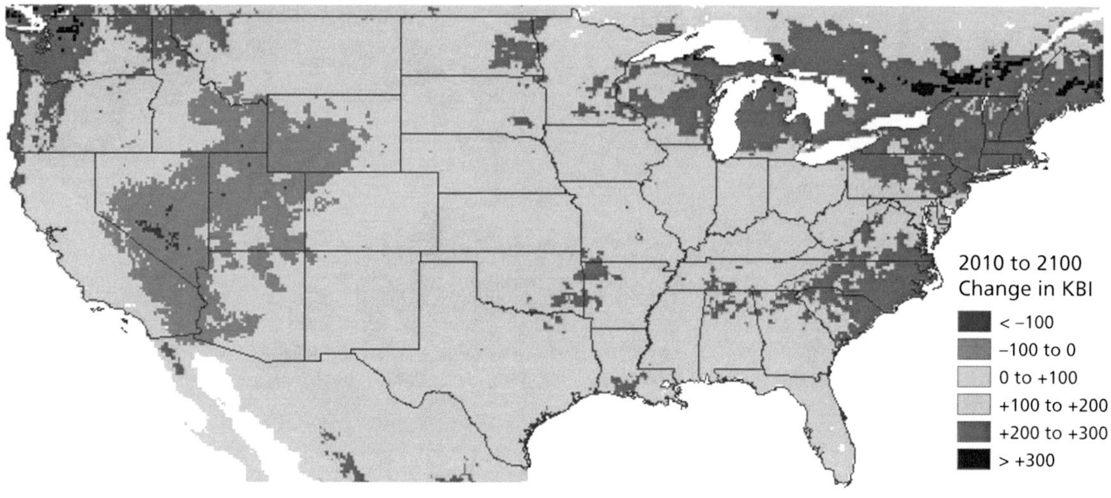


NOTE: Calculated as change in 75th percentile KBDI values for RCP 8.5 between 2006–2020 and 2090–2099.
RAND *RR1453/1-2.8*

[9] For this adjustment, we assume that multiplication factors can be calculated by equating levels of the log-normally distributed Wildland Fire Potential to levels of the KBDI, where each level of KBDI is assumed to be 200 units as in Liu, Stanturf, and Goodrick (2010). More precisely, an adjustment factor to the Wildland Fire Potential is calculated based on the change of KBDI as:

$$e^{\frac{\Delta KBDI}{200}}.$$

**Table 2.11**
**Estimation of Climate-Induced Changes to Wildfire Fire Potential**

| Analysis Factor | Description |
|---|---|
| Years | 2006–2020 (to represent current conditions)<br>2035–2045 (to represent conditions in 2040)<br>2060–2070 (to represent conditions in 2065)<br>2090–2100 (to represent the end of the century, i.e., 2100) |
| Data | CMIP5 Climate Data, 1/8 degree<br>• Daily maximum temperature<br>• Daily precipitation |
| Scenarios | • RCP 4.5<br>• RCP 8.5 |
| Return periods | N/A |
| Models | Projections will be performed for each model separately because of the variance in temperature projections across models:<br>• RCP 4.5: 22 models<br>• RCP 8.5: 22 models |
| Frequency (of drought adjustment) | • 25th percentile<br>• 50th percentile (median)<br>• 75th percentile<br>• 95th percentile<br>• Mean<br>• Max |
| Risk thresholds | • High (adjusted WFP above 401)<br>• Very High (adjusted WFP above 1935) |
| Model aggregation | • 25th percentile of updated wildfire index across model runs<br>• Median (50th percentile) of updated wildfire index across model runs<br>• 75th percentile of updated wildfire index across model runs<br>• 95th percentile of updated wildfire index across model runs<br>• Mean of updated wildfire index across model runs<br>• Maximum of updated wildfire index across model runs |
| Procedure | For each combination of 1/8 degree cell and scenario/model, the following procedures were performed:<br>1. Use the KBDI statistics produced in steps 1 and 2 of the dryness climate modeling.<br>2. Adjust the Wildland Fire Potential map using the following equation: $e^{\frac{\Delta KBDI}{200}}$.<br>3. ΔKBDI is calculated as = the difference between time periods and the baseline (e.g., 2100–2015) for each individual climate model (e.g., CanESM2_r1i1p1), for each emissions scenario (e.g., RCP 4.5), and for each frequency metric (e.g., Q75 ) compare future Wildland Fire Potential with risk thresholds and locations of infrastructure. |

NOTE: Frequency of drought adjustment reflects the percentage of time in a year the KBDI would be expected to be at or lower than the stated percentile.

**Figure 2.9**
**Wildland Fire Potential Adjusted with CanESM2 r2i1p1 Model**

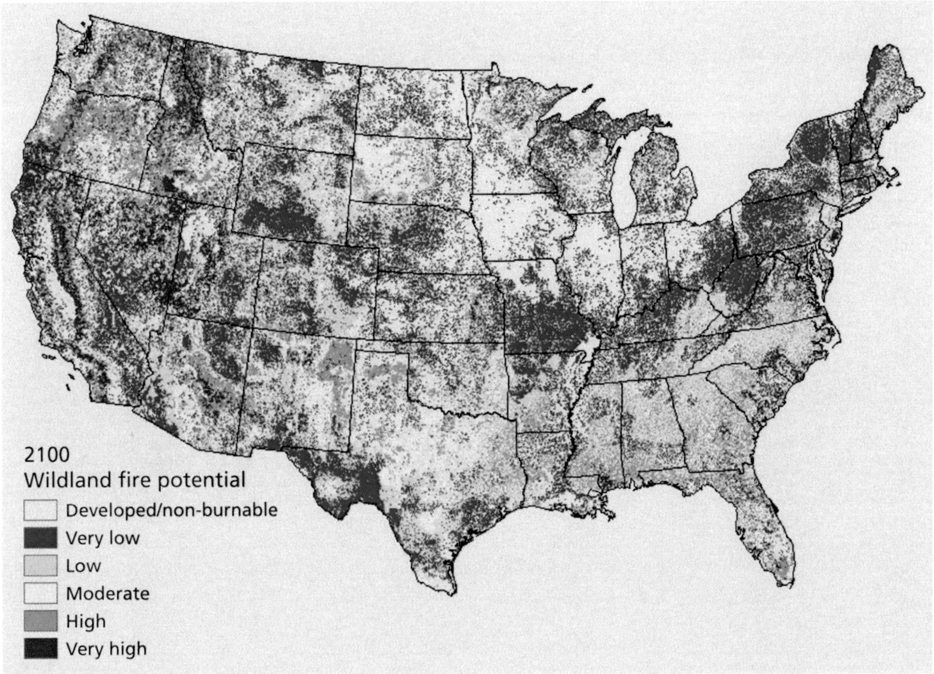


NOTE: Adjustment based on change in 75th percentile KBDI values for RCP 8.5 between 2005–2015 and 2090–2099.
RAND *RR1453/1-2.9*

Figure 2.9 shows the Wildland Fire Potential adjusted for 2100 based on the changes in the KBDI projected by the CanESM2 r2i1p1 model. Comparing Figure 2.7 reveals many places where fire potential increases significantly. The biggest increases are in the Northwest, the Southeast, and around Texas, but many other areas also increase significantly. Very few areas show a noticeable decrease in fire potential.

### 2.4.3. References

Banks, Sam C., Michaela D. J. Blyton, David Blair, Lachlan McBurney, and David B. Lindenmayer, "Adaptive Responses and Disruptive Effects: How Major Wildfire Influences Kinship-Based Social Interactions in a Forest Marsupial," *Molecular Ecology*, Vol. 21, No. 3, 2012, pp. 673–684.

Gamble, Janet L., Kristie L. Ebi, Anne E. Grambsch, Frances G. Sussman, and Thomas J. Wibanks, *Analyses of the Effects of Global Change on Human Health and Welfare and Human Systems*, Synthesis and Assessment Product 4.6, Final Report by the U.S. Climate Change Science Program and the Subcommittee on Global Change Research, U.S. Environmental Protection Agency, 2008.

Dillon, Greg, Jim Menakis, and Frank Fay, "Mapping Wildland Fire Potential for the Coterminous United States," United States Department of Agriculture, Forest Service, Fire Modeling Institute, 2013.

Finney, Mark A., Charles W. McHugh, Isaac C. Grenfell, Karin L. Riley, and Karen C. Short, "A Simulation of Probabilistic Wildfire Risk Components for the Continental United States," *Stochastic Environmental Research and Risk Assessment*, Vol. 25, 2011, pp. 973–1000.

Handmer, John, Yasushi Honda, Zbigniew W. Kundzewicz, Nigel Arnell, Gerardo Benito, Jerry Hatfield, Ismail Fadl Mohamed, Pascal Peduzzi, Shaohong Wu, Boris Sherstyukov, Kiyoshi Takahashi, and Zheng Yan, "Changes in Impacts of Climate Extremes: Human Systems and Ecosystems," in Intergovernmental Panel on Climate Change, eds., Managing the Risks of Extreme Events and Disasters to Advance Climate Change Adaption (SREX), New York: Cambridge University Press, 2012, pp. 231–290.

Keetch, John J., and George Byram, *A Drought Index for Forest Fire Control*, research paper SE-38, Asheville, N.C.: U.S. Department of Agriculture, Forest Service, Southeastern Forest Experiment Station, 1968. As of April 12, 2016:
http://www.srs.fs.usda.gov/pubs/viewpub.jsp?index=40

———, *A Drought Index for Forest Fire Control*, research paper SE-273, Asheville, N.C.: U.S. Department of Agriculture, Forest Service, Southeastern Forest Experiment Station, November 1988. As of April 12, 2016:
http://www.srs.fs.fed.us/pubs/rp/rp_se273.pdf

Liu, Yongqiang, Scott L. Goodrick, and John A. Stanturf, "Future U.S. Wildfire Potential Trends Projected Using a Dynamically Downscaled Climate Change Scenario," *Forest Ecology and Management*, Vol. 294, 2013, pp. 120–135.

Liu, Yongqiang, John Stanturf, and Scott Goodrick, "Trends in Global Wildfire Potential in a Changing Climate," *Forest Ecology and Management*, Vol. 259, 2010, pp. 685–697.

McMichael, A. J., D. H. Campbell-Lendrum, K. Ebi, A. Githeko, J. Scheraga and A. Woodward, *Climate Change and Human Health: Risks and Responses*, Geneva: World Health Organization, 2003.

Morton, D. C., M. E. Roessing, A. E. Camp and M. L. Tyrrell, *Assessing the Environmental, Social, and Economic Impacts of Wildfire*, New Haven, Conn.: Yale University, School of Forestry and Environmental Studies, Global Institute of Sustainable Forestry, 2003.

Murray, Virginia, and Kristie L. Ebi, "IPCC Special Report on Managing the Risks of Extreme Events and Disasters to Advance Climate Change Adaptation (SREX)" *Journal of Epidemiology and Community Health*, Vol. 66, No. 9, 2012, pp. 759–760.

Thompson, Matthew P., David E. Calkin, Mark A. Finney, Alan A. Ager, and Julie W. Gilbertson-Day, "Integrated National-Scale Assessment of Wildfire Risk to Human and Ecological Values," *Stochastic Environmental Risk Research and Risk Assessment*, Vol. 25, 2011, pp. 761–780.

USDA—*See* U.S. Department of Agriculture.

U.S. Department of Agriculture, "Wildfire Hazard Potential," U.S. Forest Service, Fire Modeling Institute, Fire, Fuel, and Smoke Science Program, modified October 30, 2015. As of April 12, 2016:
http://www.firelab.org/fmi/data-products/229-wildland-fire-potential-wfp

CHAPTER THREE

# Hazards Without Climate Adjustment

This chapter describes the data and methods used to analyze the following five hazards that are not adjusted for climate change:

- earthquakes
- hurricane winds
- ice storms
- riverine flooding
- tsunamis
- tornadoes
- landslides.

As described previously, climate change could affect the distribution of exposure to several of these hazards. For example, patterns of occurrence for hurricane winds, ice storms, riverine flooding, tornadoes, and landslides could all be influenced by changes in weather and climate. However, these effects are not included in this analysis because literature on these hazards does not conclusively describe how these distributions would change.

As in Chapter Two, for each hazard, we provide a brief overview of the significance of the hazard and describe the specific data sets and analysis methods used to assess the likelihood of regional exposure to varying levels of hazard severity.

## 3.1. Earthquakes

Earthquakes can have serious economic, social, and environmental impacts. Several factors, including characteristics of the earthquake itself (e.g., where it falls on the Richter scale) and characteristics of the affected region (e.g., population density and infrastructure clustering), govern the impact of an earthquake. From an infrastructure standpoint, in the short term, earthquakes could destroy infrastructure above and below the ground. Buildings and transit systems could be destroyed. Underground gas pipelines could explode, leading to fires and related secondary effects. In the longer term, the economic and social costs of rebuilding infrastructure following an earthquake could be enormous (American Society of Civil Engineers, 2007).

An understanding of the expected likelihood and severity of earthquakes affecting the region informs investments targeted at boosting the resilience of a community or area to seismic activity.

### 3.1.1. Data Sources

Data for seismic hazards are taken from the USGS Seismic Hazards layer contained in the HSIP Gold data set. Associated with the data are 500- and 2,500-year return periods.

### 3.1.2. Analysis Methods

The USGS National Seismic Hazard Mapping Project creates earthquake hazard maps for the United States intended to guide building codes (USGS, undated). The USGS and its partners create the maps using knowledge of historical earthquakes and geology. The current maps were made in 2008, but periodic updates occur as building codes are revised, and updates incorporate new research results (USGS, 2008).

The USGS maps include two levels of likelihood (–2-percent or 10-percent chance of being exceeded in a 50-year period) and three types of consequences (peak ground acceleration [PGA], peak ground velocity, and spectral acceleration). The likelihood percentages correspond roughly to a 0.04-percent AEP (or a return period of about 2,500 years) and a 0.2-percent AEP (or a return period of around 500 years).[1]

In this analysis, we use PGA because it is a standard determinant of infrastructure damage. USGS maps bin PGA into seven ranges: 0–4 percent, 4–8 percent, 8–16 percent, 16–32 percent, 32–48 percent, 48–64 percent, and 64 percent plus gravitational acceleration (g).

Our analysis sets thresholds on PGA and return periods to classify earthquakes by intensity. Earthquakes are classified according to their severity corresponding to the Modified Mercalli Intensity Scale in the following way:

- low severity–observed damage: PGA above 0.1 g
- high severity–severe shaking: PGA above 0.5 g.

These maps are available for the entire United States, but they are relatively coarse. Detailed studies—for example, of specific metro areas—can account for more detailed spatial differences, such as soil types. However, such maps are not available for the entire country. Figures 3.1 and 3.2 show sample maps for the Los Angeles area and CONUS, respectively, indicating the PGA associated with a 2,500-return period (or 0.0-percent AEP).

Climate change is not expected to increase earthquake risk significantly, so we do not include it in our climate-adjusted hazards. The SREX does cite instances where climate change can be expected to increase the frequency of small earthquakes, for example, as glaciers recede (Seneviratne et al., 2012, p. 188), but these earthquakes are small and tend to be in isolated regions.

---

[1] Exact return periods are 2,475.9 years and 474.6 years, respectively (USGS, 2016).

**Figure 3.1**
**2-Percent PGA for the 0.04 Percent AEP in the Los Angeles Area**

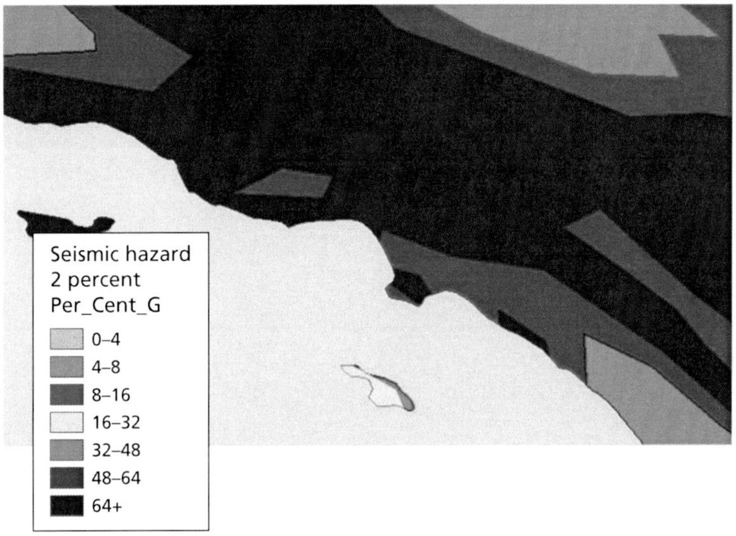


RAND *RR1453/1-3.1*

**Figure 3.2**
**2-Percent PGA for the 0.04 Percent AEP in CONUS**

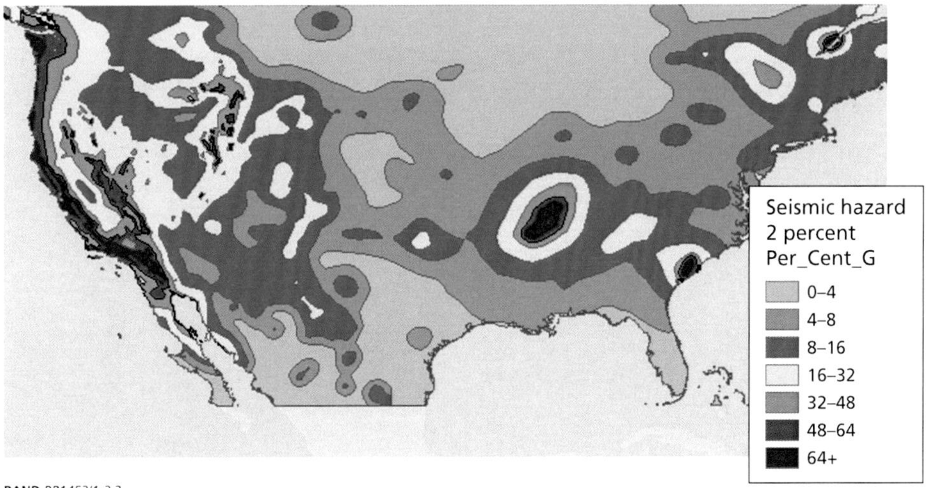


RAND *RR1453/1-3.2*

### 3.1.3. References

American Society of Civil Engineers, Seismic Rehabilitation of Existing Buildings, ASCE 41-06, Reston, Va., 2007.

Seneviratne, Sonia, Neville Nicholls, David Easterling, Clare M. Goodess, Shinjiro Kanae, James Kossin, Yali Luo, Jose Marengo, Kathleen McInnes, Mohammad Rahimi, Markus Reichstein, Asgeir Sorteberg, Carolina Vera, and Xuebin Zhang, "Changes in Climate Extremes and Their Impacts on the Natural Physical Environment," in Intergovernmental Panel on Climate Change, eds., *Managing the Risks of Extreme Events and Disasters to Advance Climate Change Adaptation (SREX)*, New York: Cambridge University Press, 2012, pp. 109–230.

USGS—*See* U.S. Geological Survey.

U.S. Geological Survey, "National Seismic Hazard Mapping Project," U.S. Department of the Interior, web page, undated. As of April 14, 2016:
http://earthquake.usgs.gov/hazards/

———, "United States National Seismic Hazards Maps," Fact Sheet 2008-3017, U.S. Department of the Interior, April 2008. As of April 14, 2016:
http://pubs.usgs.gov/fs/2008/3017/pdf/FS08-3017_508.pdf

———, "Earthquake Hazards 201—Technical Q&A," U.S. Department of the Interior, web page, revised April 6, 2016. As of April 14, 2016:
http://earthquake.usgs.gov/hazards/about/technical.php

## 3.2. Hurricane Winds

Hurricane winds can topple power lines, destroy buildings, bring down trees, and create dangerous flying debris. The higher the wind speeds associated with a hurricane, the greater the potential for property damage. Tropical cyclones are commonly associated with extreme winds; the extent of a tropical cyclone's wind field can play a significant role in determining the impacts from storm surge. Like other extreme phenomena, wind extremes can by defined by percentiles, maximums over a stated time period, or a threshold value (e.g., 99th percentile 10-meter hourly wind). Wind gusts are measured by the highest winds during a short time period. However, extreme wind is often not characterized from its own observations, but rather defined by the hazard phenomena that generated it (including tropical and extratropical cyclones and thunderstorms). In turn, wind speed can influence additional climate stressors including SLR, wildfires and droughts.

Extreme wind changes need to be distinguished from natural wind variability flows. For example, there is substantial tropical cyclone frequency variability based on different regions and ocean basins (Karl et al., 2008; Murray and Ebi, 2012). In this analysis we focus on the effects of high wind speeds that typically accompany hurricanes.

The IPCC's SREX reviewed literature on historical tropical cyclones and found that there is low confidence in trends in historical frequency (Seneviratne et al., 2012, p. 163). The SREX also concluded that there is low confidence in how the geography of tropic cyclones (i.e., the tracks and areas of impact) will change in the future. The SREX states that it is likely that the frequency of tropical cyclones will not increase globally, even though it is likely that rainfall rates and maximum wind speeds will increase in tropical cyclones. The SREX states with high confidence that the damages from tropical cyclones will increase from increases in exposure (Handmer et al., 2012, p. 271), but that the main driver for increasing losses in many regions will likely be socioeconomic factors rather than climate change–related factors (Handmer et al., 2012, p. 273). Still, these effects could modify exposures to extreme winds. These effects were not considered in the climate scenarios described here. However, this literature does not describe how the distribution of exposure will be modified. Depending on where exposures are occurring, the resulting analysis either underestimates or overestimates exposures.

### 3.2.1. Data Sources

We use hurricane wind speed data included in the HAZUS Hurricane Model (MH 2.1) that contain return periods of 20, 50, 100, 200, 500 and 1,000 years.

### 3.2.2. Analysis Methods

The data used in this study consider different wind-speed return periods: 10, 20, 50, 100, 200, 500 and 1,000 years. We use two maximum wind-speed thresholds to classify the severity of different hurricane scenarios. Hurricane scenarios in which maximum wind speeds exceed 96 miles per hour are classified as low-severity hurricanes; wind speeds above 130 miles per hour represent high-severity hurricanes. These cut points are in agreement with the HAZUS scale, as shown in Figure 3.3 for a sample return period of 100 years.

**Figure 3.3**
**Hurricane Winds, Present-Day 100-Year Return Wind Speeds**

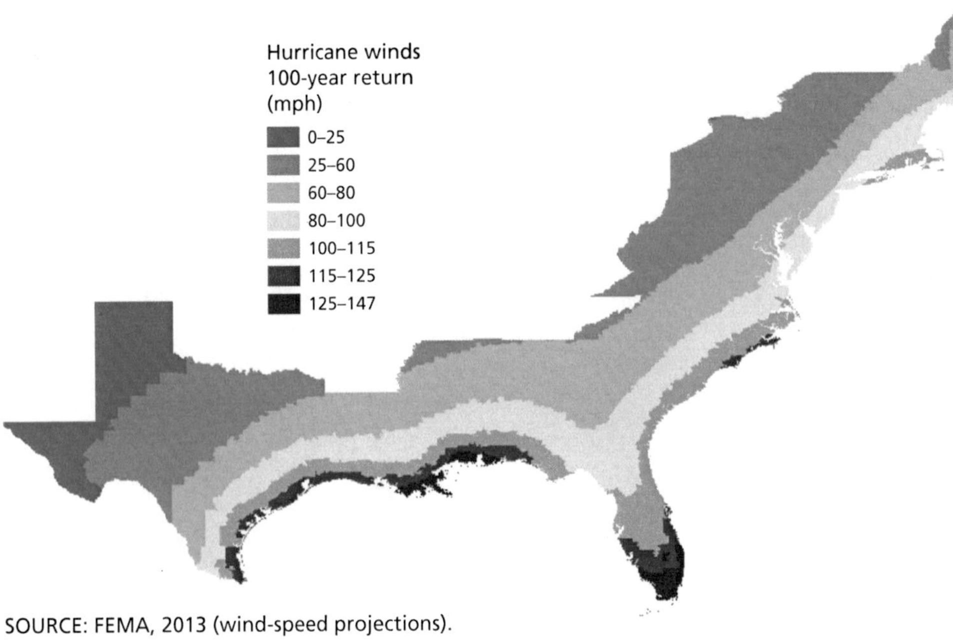


SOURCE: FEMA, 2013 (wind-speed projections).
RAND *RR1453/1-3.3*

### 3.2.3. References

Federal Emergency Management Agency, "Hazus-MH 2.1, Hurricane Model, Technical Manual," 2013. As of April 14, 2016:
http://www.fema.gov/media-library-data/20130726-1820-25045-9850/hzmh2_1_hr_tm.pdf

FEMA—*See* Federal Emergency Management Agency.

Handmer, John, Yasushi Honda, Zbigniew W. Kundzewicz, Nigel Arnell, Gerardo Benito, Jerry Hatfield, Ismail Fadl Mohamed, Pascal Peduzzi, Shaohong Wu, Boris Sherstyukov, Kiyoshi Takahashi, and Zheng Yan, "Changes in Impacts of Climate Extremes: Human Systems and Ecosystems," in Intergovernmental Panel on Climate Change, eds., *Managing the Risks of Extreme Events and Disasters to Advance Climate Change Adaptation (SREX)*, New York: Cambridge University Press, 2012, pp. 231–290.

Karl, Thomas, Gerald A. Meehl, Christopher D. Miller, Susan J. Hassol, Anne M. Waple, and William L. Murray, *Weather and Climate Extremes in a Changing Climate—Regions of Focus: North America, Hawaii, Caribbean, and U.S. Pacific Islands*, Synthesis and Assessment Product 3.3, U.S. Climate Change Science Program, June 2008.

Murray, Virginia, and Kristie L. Ebi, "IPCC Special Report on Managing the Risks of Extreme Events and Disasters to Advance Climate Change Adaptation (SREX)," *Journal of Epidemiology and Community Health*, Vol. 66, No. 9, September 2012, pp. 759–760.

Seneviratne, Sonia, Neville Nicholls, David Easterling, Clare M. Goodess, Shinjiro Kanae, James Kossin, Yali Luo, Jose Marengo, Kathleen McInnes, Mohammad Rahimi, Markus Reichstein, Asgeir Sorteberg, Carolina Vera, and Xuebin Zhang, "Changes in Climate Extremes and Their Impacts on the Natural Physical Environment," in Intergovernmental Panel on Climate Change, eds., *Managing the Risks of Extreme Events and Disasters to Advance Climate Change Adaptation (SREX)*, New York: Cambridge University Press, 2012, pp. 109–230.

## 3.3. Ice Storms

Following an extreme ice storm in 1998 that led to widespread, long-term power outages in Quebec, Canada, and adjacent U.S. states, there was an increased recognition of the potential damage from winter ice storms to infrastructure. In response, FEMA formed the American Lifelines Alliance, which in 2004 released a study that estimated the likelihood and magnitude of winter storms (American Lifeline Alliance, 2004). Deposits of heavy amounts of ice from ice storms can damage infrastructure. Wind further exacerbates the force of the ice, thus the estimates look at both ice thickness and the winds that would accompany or follow the ice storm. To produce these estimates, the researchers used their expert judgment along with newspaper and weather reports since 1950.

Ice storms likely will be impacted by climate change, but not much research exists on what these changes will be. Historically in the United States, there has been a trend for more extreme snowstorms—even in warm years—but no trends in ice storms have been observed over the past century (Kunkel et al., 2013). Our inclusion of ice storms, but not snowstorms, reflects the fact that ice storms can damage infrastructure, whereas snowstorms are more likely to merely disrupt infrastructure.

### 3.3.1. Data Sources

Ice storm data were taken from the digitized data available from the American Lifelines Alliance's *Extreme Ice Thickness from Freezing Rain* hazard assessment (American Lifelines Alliance, 2004). We consider a 50-year return period for ice storms.

### 3.3.2. Analysis Methods

The American Lifelines Alliance study estimated four return periods for ice storms (50, 100, 200, and 400 years). A parallel, ongoing effort estimates the occurrence of extreme ice storms for use in building codes, contained in ASCE 7-10 (American Society of Civil Engineers, 2013). ASCE 7-10 uses the 50-year return maps. We obtained a digitized version of the 50-year return maps directly from the USACE, Cold Regions Research and Engineering Laboratory (Jones, 2014).[2] Other returns could be used, but they would require digitizing maps from the 2004 American Lifelines Alliance study.

We used cutoffs for estimating potential ice damage using the ice thickness and wind speed projections defined by the Sperry-Piltz Ice Accumulation (SPIA) Index (see Figure 3.4). The index ranges from 0 to 5, with higher levels indicating larger ice loads, higher wind speeds, and greater potential for damage to electrical power systems.

Figure 3.5 shows the SPIA index that is calculated for a 50-year return period for CONUS. Note that a large portion of the country is in the most severe category (5), meaning that a catastrophic ice storm is expected at least once every 50 years. Additionally, the hatching in the figure indicates mountainous areas with topography that could lead to more severe ice than is predicted by the lower-fidelity analysis used to create the projections. This means that small areas can have a more severe SPIA index for the 50-year return period than designated by the index value for the surrounding area.

---

[2] The digital maps we obtained were publication quality, but they needed some additional adjustments to be used in our analysis. We adjusted the lines so that they were continuous and complete, which required minor shifts in the lines we received from the Cold Regions Research and Engineering Laboratory.

**Figure 3.4**
**The Sperry-Piltz Ice Accumulation Index**

| Ice Damage Index | *Average NWS Ice Amount (inches) *Revised—October 2011 | Wind (mph) | Damage and Impact Descriptions |
|---|---|---|---|
| 0 | < 0.25 | < 15 | Minimal risk of damage to exposed utility systems; no alerts or advisories needed for crews; few outages. |
| 1 | 0.10–0.25 | 15–25 | Some isolated or localized utility interruptions are possible, typically lasting only a few hours. Roads and bridges may become slick and hazardous. |
| | 0.25–0.50 | < 15 | |
| 2 | 0.10–0.25 | 25–35 | Scattered utility interruptions expected, typically lasting 12 to 24 hours. Roads and travel conditions may be extremely hazardous due to ice accumulation. |
| | 0.25–0.50 | 15–25 | |
| | 0.50–0.75 | < 15 | |
| 3 | 0.10–0.25 | ≥ 35 | Numerous utility interruptions with some damage to main feeder lines and equipment expected. Tree limb damage is excessive. Outages lasting 1 to 5 days. |
| | 0.25–0.50 | 25–35 | |
| | 0.50–0.75 | 15–25 | |
| | 0.75–1.00 | < 15 | |
| 4 | 0.25–0.50 | ≥ 35 | Prolonged and widespread utility interruptions with extensive damage to main distribution feeder lines and some high voltage transmission lines/structures. Outage lasting 5 to 10 days. |
| | 0.50–0.75 | 25–35 | |
| | 0.75–1.00 | 15–25 | |
| | 1.00–1.50 | < 15 | |
| 5 | 0.50–0.75 | ≥ 35 | Catastrophic damage to entire exposed utility systems, including both distribution and transmission networks. Outages could last several weeks in some areas. Shelters needed. |
| | 0.75–1.00 | ≥ 25 | |
| | 1.00–1.50 | ≥ 15 | |
| | > 1.50 | Any | |

SOURCE: Sperry-Piltz Ice Accumulation Index, web page, February 2009.
NOTE: Categories of damage are based upon combinations of precipitation totals, temperatures, and wind speeds/directions.

RAND *RR1453/1-3.4*

**Figure 3.5**
**The Sperry-Piltz Ice Accumulation Index Applied to CONUS**

SOURCE: RAND calculations applying the SPIA index to the 50-year return ice thickness and wind from American Lifelines Alliance, 2004.

RAND *RR1453/1-3.5*

Because so much of the country has a relatively high likelihood (2 percent or more each year) of catastrophic ice damage, more-refined modeling can focus only on areas with an index of 5 and differentiate those areas by ice thickness, see Figure 3.6.

**Figure 3.6**
**Ice Thickness for 50-Year Return Period for Areas with a Sperry-Piltz Ice Accumulation Index of 5**

SOURCE: RAND calculations applying the SPIA index to the 50-year return ice thickness and wind from American Lifelines Alliance, 2004.
RAND *RR1453/1-3.6*

### 3.3.3. References

American Lifelines Alliance, *Extreme Ice Thickness from Freezing Rain*, Washington, D.C., September 2004. As of April 14, 2016:
http://americanlifelinesalliance.com/pdf/CombinedALAFinalReport092804.pdf

American Society of Civil Engineers, *Minimum Design Loads for Buildings and Other Structures*, ASCE 7-10, Third Printing, Reston, Va., 2013.

Jones, Kathleen D., "Shapefiles for Extreme Ice Thickness from Freezing Rain for 50-Year Return Period," email with authors, January 16, 2014.

Kunkel, Kenneth E., Thomas R. Karl, Harold Brooks, James Kossin, Jay H. Lawrimore, Derek Arndt, Lance Bosart, David Changnon, Susan L. Cutter, Nolan Doesken, Kerry Emanuel, Pavel Ya. Groisman, Richard W. Katz, Thomas Knutson, James O'Brien, Christopher J. Paciorek, Thomas C. Peterson, Kelly Redmond, David Robinson, Jeff Trapp, Russell Vose, Scott Weaver, Michael Wehner, Klaus Wolter, and Donald Wuebbles, "Monitoring and Understanding Trends in Extreme Storms: State of Knowledge," *Bulletin of the American Meteorological Society*, Vol. 94, No. 4, April 2013, pp. 499–514.

Sperry-Piltz Ice Accumulation Index, web page, February 2009. As of April 15, 2016:
http://www.spia-index.com/

## 3.4. Riverine Flooding

Riverine flooding is an extremely relevant natural hazard, as many population centers have developed around rivers and communities have been established inside floodplains that are currently dry but could be reclaimed by water in future. As a result, riverine flooding can be a costly natural hazard (Olsen, 2006). For instance, the 1993 summer riverine flooding in the upper Mississippi river was a devastating flood event that affected nine states with estimated economic losses up to $20 billion (Kunkel, Changnon, and Angel, 1993).

Riverine flooding occurs when flows going into natural water bodies, such as rivers, streams, and lakes exceed the capacity of these banks, causing overflows to adjacent areas. Riverine flooding can be induced by excessive runoff from intense rainfall, channel erosion, and infrastructure failure (i.e., dams and levees).

There are two main types of riverine flooding (Emergency Management Institute, 2007):

1. **Overbank flooding:** This is the overflow of water from river channels and stream flows into adjacent areas. The magnitude and consequences of these events vary considerably. In the United States, hundreds of events are registered annually.
2. **Flash floods:** These types of riverine flood events are characterized by a sudden rise in water, high flow velocities, and large volumes of debris. This type of riverine flooding occurs across the entire United States, but most commonly in mountainous regions. The damage from this type of riverine flooding can be quite severe on infrastructures or populations in the vicinity of rivers. Streamflow velocities of 9 feet per second are commonly registered in these types of events. However, streamflow velocities can be as high as 30 feet per second. Water moving at these velocities can have enormous impacts on infrastructures and populations. For instance, water bodies moving at speeds of 15 feet per second are capable of washing out access roads, which can have severe negative effects on the authorities' capacities to respond to an emergency, putting human lives at risk. The NWS has identified approximately 2,000 communities across CONUS in which flash floods are a risk.

FEMA classifies areas at risk of riverine flooding using the SFHA classification system. SFHAs are areas subject to a 100-year return period flooding. Riverine flooding falls mainly into two categories: A or AE zones. FEMA defines A zones as "areas subject to inundation by the 1-percent-annual-chance flood event generally determined using approximate methodologies" and AE zones as "areas subject to inundation by the 1-percent-annual-chance flood event determined by detailed methods" (FEMA, 2015a). The difference between these zones is that the chance and impact of flooding are estimated using different methods. For AE areas, precise flood levels are determined, while only approximate levels of flooding are provided for A zones. Regardless, these classifications are determined using numerical models for riverine flood analyses. These methods include the Hydrologic Engineering Center Modeling System (HEC-HMS) and the U.S. Geological Survey National Flood Frequency program (Crowell et al., 2010).

Climate change is expected to influence the frequency and intensity of riverine floods because of changes in precipitation patterns and other climate-influenced factors (Walsh et al., 2014). We are aware of at least one preliminary study looking at the potential effects of climate change on future riverine flooding across the United States (AECOM, Michael Baker Jr. Inc.

and LLP Deloitte Consulting, 2013), and recently routed CMIP5 river flows could be adapted for a future national assessment of climate-influenced riverine flood exposure.

At present, however, there are no data sets that systematically describe potential changes in riverine flood depths across the United States as a result of climate change, and it was determined to be beyond the scope of this analysis to develop and apply such an analysis. As a result, we consider riverine flood exposure in a simplistic manner only, and do not consider potential future changes for this hazard.

### 3.4.1. Data Sources

Riverine flooding data are derived from the FEMA SFHAs (100-year return period), provided in the National Flood Hazard Layer (FEMA, 2015b).

### 3.4.2. Analysis Methods

The primary data source for riverine flooding is FEMA's National Flood Hazard Layer, which provides FEMA's estimate of the 1 percent AEP flood elevation. As with areas exposed to coastal flooding, described in Section 2.1.2, the 1-percent AEP elevation is used to identify riverine flooding SFHAs for the NFIP's FIRMs. Residents of the SFHA are required to purchase flood insurance via NFIP under most circumstances—in order to qualify for a federally insured mortgage, for example.

The National Flood Hazard Layer includes flood maps that have been formally adopted and digitized. The actual flood elevation and flood depth data do not include all areas exposed to flood risk in the United States, however, with gaps from (a) paper maps that have not yet been digitized and (b) a slow analytic update and approval process for new maps. As a result, not all regions can be included in the analysis focused on the 100-year floodplain.

Data available to support nationwide assessments of riverine flooding are limited. In many riverine flood areas, FEMA provides the boundaries of the SFHA, but it does not provide estimates of BFEs or flood depths within these areas.

As a result, this analysis uses a very simple binary assessment to assess riverine flood exposure. Using a simple geospatial comparison, any facilities within a noncoastal SFHA polygon are counted as exposed to riverine flooding, whereas those outside the polygons boundaries are not considered. This assessment was made using a snapshot of the SFHAs obtained in 2014, and assumes no future change.

### 3.4.3. References

AECOM, Michael Baker Jr. Inc., and LLP Deloitte Consulting, *The Impact of Climate Change and Population Growth on the National Flood Insurance Program Through 2100*, prepared for Federal Insurance and Mitigation Administration and Federal Emergency Management Agency, June 2013. As of March 24, 2015:
http://www.aecom.ca/vgn-ext-templating/v/index.jsp?vgnextoid=e0642ed99724e310VgnVCM100000089e1bacRCRD

Crowell, Mark, Kevin Coulton, Cheryl Johnson, Jonathan Westcott, Doug Bellomo, Scott Edelman, and Emily Hirsch, "An Estimate of the U.S. Population Living in 100-Year Coastal Flood Hazard Areas," *Journal of Coastal Research*, Vol. 26, No. 2, 2010, pp. 201–211.

Emergency Management Institute, "Chapter 2: Types of Floods and Floodplains," in James M. Wright, eds., *Floodplain Management: Principles and Current Practices*, Emmitsburg, Md.: Federal Emergency Management Agency, 2007, pp. 2-1–2-11. As of June 3, 2016:
https://training.fema.gov/hiedu/docs/fmc/chapter%202%20-%20types%20of%20floods%20and%20floodplains.pdf

Federal Emergency Management Agency, "Flood Zones," web page, April 26, 2015a. As of April 14, 2016: https://www.fema.gov/flood-zones

———, "National Flood Hazard Layer (NFHL)," web page, August 21, 2015b. As of March 24, 2015: https://www.fema.gov/national-flood-insurance-program-flood-hazard-mapping/national-flood-hazard-layer-nfhl#

FEMA—*See* Federal Emergency Management Agency.

Kunkel, Kenneth E., Stanley A. Changnon, and James R. Angel, "Climatic Aspects of the 1993 Upper Mississippi River Basin Flood," *Bulletin of the American Meteorological Society*, Vol. 75, No. 5, 1994, pp. 811–822.

Olsen, J. Rolf, "Climate Change and Floodplain Management in the United States," *Climatic Change*, Vol. 76, No. 3–4, 2006, pp. 407–426.

Walsh, John, Donald Wuebbles, Katharine Hayhoe, James Kossin, Kenneth Kunkel, Graeme Stephens, Peter Thorne, Russell Vose, Michael Wehner, Josh Willis, David Anderson, Scott Doney, Richard Feely, Paula Hennon, Viatcheslav Kharin, Thomas Knutson, Felix Landerer, Tim Lenton, John Kennedy, and Richard Somerville, "Ch. 2: Our Changing Climate," in Jerry M. Melillo, Terese (T. C.) Richmond, and Gary W. Yohe, eds., *Climate Change Impacts in the United States: The Third National Climate Assessment, U.S. Global Change Research Program*, 2014, pp. 19–67. As of March 24, 2015: http://nca2014.globalchange.gov/report/our-changing-climate/introduction

## 3.5. Tsunamis

Tsunamis can have a range of impacts from barely being noticed to causing widespread and devastating damage. As with earthquakes, the extent of a tsunami's impact depends on several factors including its distance from the point of origin, its magnitude, and ocean depth in the affected coastal areas. In extreme cases, tsunamis can destroy buildings, bridges, transit systems, power lines, and most structures on their path. Again, as with earthquakes, the period of recovery and rebuilding following a tsunami can be extremely taxing both economically and socially.

Another mechanism through which tsunamis can damage infrastructure is run-up, i.e., water from the force of tsunami waves pushed inland and to higher elevations. These effects are extremely dependent upon the directionality and force of waves, geography of coasts, and elevations of land. These effects are not represented in the tsunami hazard data, because detailed modeling of tsunami run-up is not available for CONUS. Thus, the exposures to tsunami hazards are likely an underestimate of exposure.

The cause of most tsunamis is not related to climate, thus they are not studied in depth by the climate literature. Climate change is likely to have some impact on tsunamis because of increases in sea level, but these are relatively small compared with the size of a tsunami. Furthermore, much of the coast in the Pacific Northwest is rising, thus mitigating some of the impact of SLR (U.S. Army Corps of Engineers, 2014).

Overall, the quality of tsunami hazard data for the United States is relatively poor. Even for regions that have studied these hazards in depth, such as Oregon (Priest, 1995), the hazard maps tend to be scenario-based instead of likelihood-based, so it is difficult to understand tsunami risk.

### 3.5.1. Data Sources

Tsunami exposure is derived from the Risk Management Solutions (RMS) Tsunami Report and USGS NED. The associated return period is ≤ 500 years.

### 3.5.2. Analysis Methods

Data supporting the assessment of tsunami risks are generally sparse and of poor quality. Better data would help strengthen the associated analysis. Recent efforts are attempting to engage in more comprehensive modeling of tsunami hazards across the United States. NOAA is developing DEMs of the ocean floor and shoreline to model tsunamis on U.S. coasts (NOAA National Geophysical Data Center, undated). These DEMs are highly detailed and can be fed into a NOAA tsunami simulator called the Method of Splitting Tsunami (MOST), which performs scenario-based simulations of tsunamis, including inundation of small areas (NOAA Center for Tsunami Research, undated).

These NOAA efforts are mainly focused on estimating, at a detailed level, the consequences of a tsunami arising from specific scenarios (e.g., an earthquake of a given magnitude occurring in a particular location). There is a gap in research that attempts to assess overall tsunami hazards, i.e., both the likelihood and the inundation of tsunamis.

There are also local and regional efforts to identify areas of high tsunami hazards. For example, the California Department of Conservation publishes tsunami hazard maps for populated coastlines (California Geological Survey, 2007). These maps were created by using the MOST model for a "suite of tsunami source events" that represents "a number of extreme, yet realistic, tsunami sources" (California Emergency Management Agency, 2009). The maps show the maximum inundation across these scenarios, but they do not contain information about the likelihood of such scenarios occurring.

RMS estimated the return period for a 5-meter tsunami along shorelines throughout the world (RMS, 2006). The RMS estimates have four bins, ranging from high hazard, where return periods are less than 500 years, to negligible hazard, where geology is not conducive to creating a tsunami. Although any coastline has some chance of being hit by a large tsunami (e.g., from a meteor strike), for the purposes of this analysis we only consider areas with a return period of less than 500 years. Therefore, we focus our tsunami hazard estimates on high-hazard areas (in red in Figure 3.7).

To identify areas that are likely to be inundated during a tsunami, we conducted a simple analysis of coastal topography on high hazard coastlines (in Northern California, Oregon, and Washington) that flags any location with an elevation of 5 meters or less as being in a high tsunami-hazard zone. The RMS hazard assessment does not differentiate between high, low, and medium hazards by tsunami height. Instead, it differentiates them only by the likelihood of a tsunami of at least 5 meters. The 5-meter height is associated with a significant loss of life—about 5 percent of the population. Tsunami heights from earthquakes can greatly exceed 5 meters and result in higher mortality rates. For example, the highest run-up in the 2004 Indian Ocean tsunami was 40 meters. However, these heights are very unlikely as they depend on specific locations of earthquakes, bathymetry, and topography. The elevation data set is from the USGS NED (USGS, 2006), which is a compilation of raster elevation data for CONUS. This analysis does not account for the effects of the ocean floor or topography, which can attenuate tsunamis. NOAA tsunami models could better account for these details, but such models would have to be run across a range of different tsunami scenarios across a range of areas, thus such modeling is not feasible in this project.

**Figure 3.7**
**RMS Tsunami Hazard Assessment for North American Coasts, 2006**

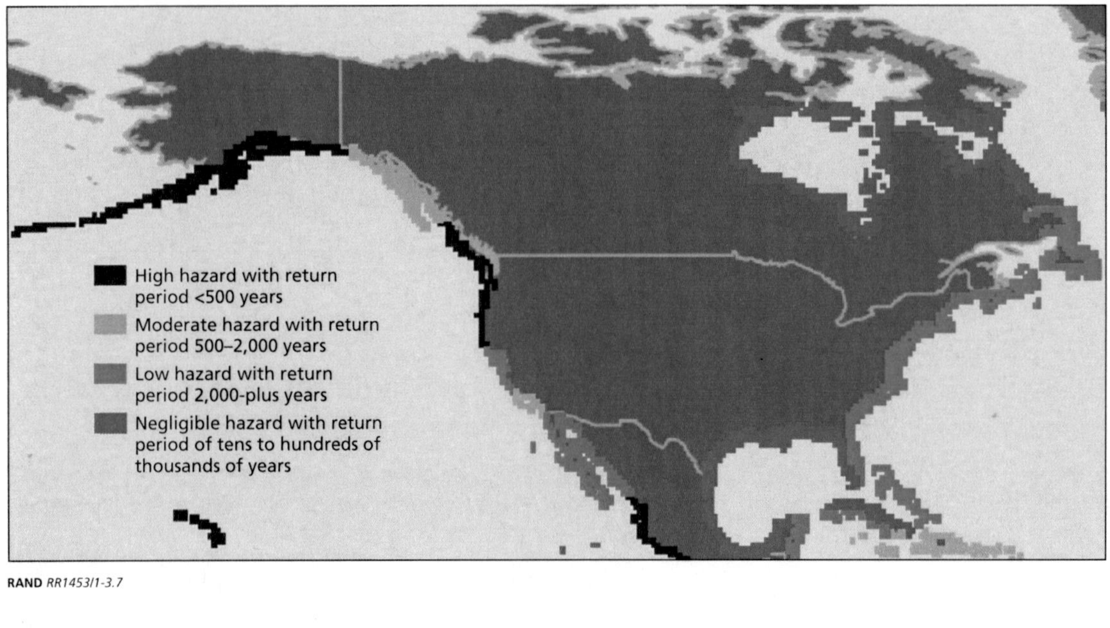


RAND *RR1453/1-3.7*

To estimate the potential area of inundation for a tsunami, the following procedure was used:

- Elevation data were selected for all areas in CONUS that are:
  - within 1 kilometer of the Pacific Coast
  - north of 40-degrees, 15-minutes north latitude (the southern limit of the RMS "high hazard" coastline)
- Any areas with an elevation at or below 5 meters were deemed at risk of a 500-year tsunami.
  - Other areas are considered outside the 5-meter tsunami zone
- To estimate the height of the tsunami, we assume that the 500-year tsunami is 5 meters tall, so the height of inundation is the difference between 5 meters and the elevation.

RMS analysis considers tsunamis that inundate land that is 0.5 to 1 kilometer from the coast. Tsunami water can reach much farther inland (for example, 4 kilometers on part of Sumatra in the 2004 tsunami); however, most of the damage occurred near the coast. The choice of 40-degrees, 15-minutes north latitude coincides with RMS's estimate of the high-hazard area with tsunami return periods of 500 years or less. These coastlines lie along subduction zones formed by the North American Plate, the Gorda Plate, and the Juan de Fuca Plate.

The tsunami estimates from RMS that we use might be conservative in some places. The high hazard zone is anything with a return period of 500 years or less. It is likely that some areas have a return period for a 5-meter tsunami of less than 500 years. In such a region, a 500-year tsunami would correspond to a tsunami above 5 meters in height. Additionally, during an earthquake that triggers a large tsunami, land can subside, which could push the tsunami run-up higher relative to the current elevation. For example, models used to construct Oregon's tsunami hazard zone project (Priest, 1995) include up to 4 feet of subsidence (see Figure 3.8).[3] Finally, if the tsunami occurred at high tide, it would run up higher (and conversely, if at low tide, it would run up lower). Unfortunately, comprehensive data that are detailed enough to make these estimates are not available.

Our method assumes that the tsunami run-up elevation is constant (within the cutoff of 1 kilometer to the coast). In reality, tsunami run-up dissipates as tsunamis travel across land, thus the run-up from a 5-meter tsunami 500 meters inland is likely to be less than 5 meters.

Figure 3.9 shows tsunami hazards predicted for Pacific City, Oregon. Note that the behavior shown on the map is similar for much of the Pacific Coast. Generally, the only area that lies below 5 meters of elevation is sandy beach (and considered on many maps to be part of the ocean). The steep beaches of the Pacific Coast mitigate the damage that would be caused by a 5-meter tsunami.

---

[3] For areas with a substantial likelihood of uplift of land, the authors assumed zero subsidence (Priest, 1995, p. 8). Subsidence was minor overall compared with the modeled tsunami heights, which reach up to 48 feet in one of their three models used to map inundation zones. Therefore, the Oregon inundation maps consider much more severe tsunamis than those used in the RMS projection.

**Figure 3.8**
**USGS National Elevation Data for Astoria, Oregon**

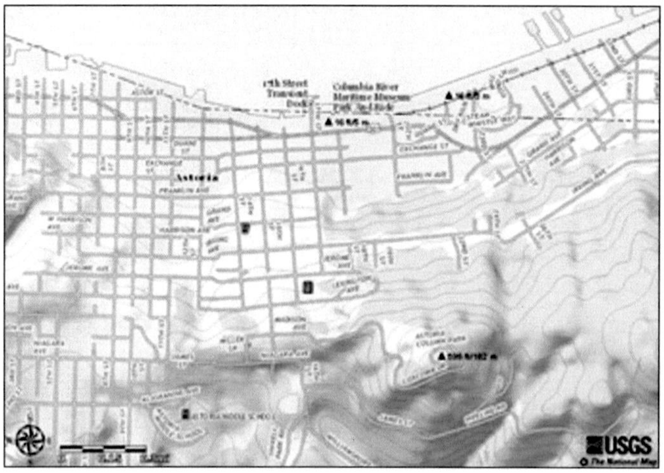


RAND *RR1453/1-3.8*

**Figure 3.9**
**RAND Estimation of Inundation Height for 5-Meter Tsunami in Pacific City, Oregon**

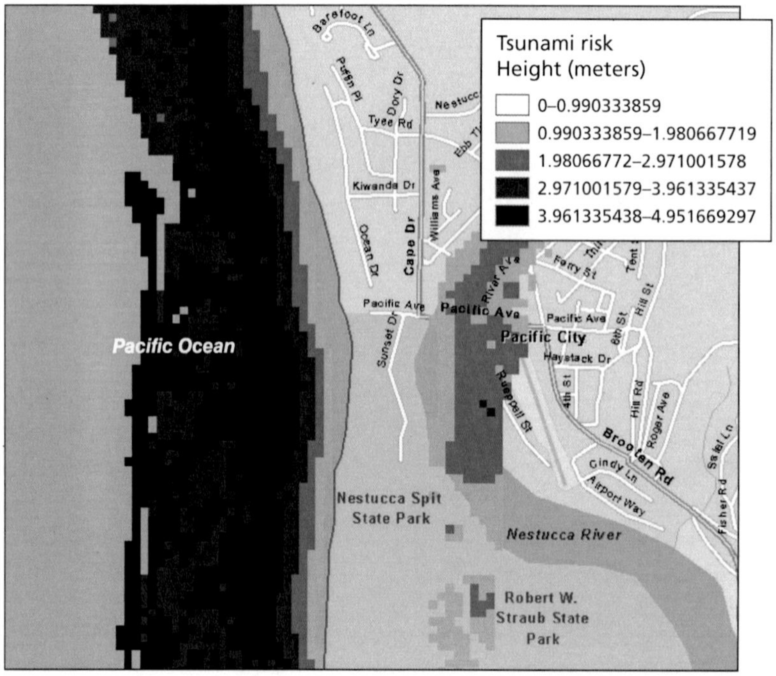


RAND *RR1453/1-3.9*

### 3.5.3. References

California Geological Survey, "CGS Information Warehouse: Tsunami," State of California Department of Conservation, 2007. As of April April 15, 2016:
http://www.quake.ca.gov/gmaps/WH/tsunamimaps.htm

California Emergency Management Agency, *Tsunami Inundation Map for Emergency Planning: Venice Quadrangle*, California Geological Survey, March 1, 2009.

National Oceanic and Atmospheric Administration Center for Tsunami Research, "Tsunami Modeling and Research," Pacific Marine Environmental Laboratory, web page, undated. As of April 15, 2016:
http://nctr.pmel.noaa.gov/model.html

National Oceanic and Atmospheric Administration National Geophysical Data Center, "NOAA Tsunami Inundation Digital Elevation Models (DEMs)," web page, undated. As of April 15, 2016:
https://www.ngdc.noaa.gov/mgg/inundation/tsunami/

NOAA—*See* National Oceanic and Atmospheric Administration.

Priest, George R., *Explanation of Mapping Methods and the Use of the Tsunami Hazard Maps of the Oregon Coast*, Open-File Report O-95-67, Portland, Ore.: State of Oregon, Department of Geology and Mineral Industries, 1995.

Risk Management Solutions, *Managing Tsunami Risk in the Aftermath of the 2004 Indian Ocean Earthquake & Tsunami*, Newark, Calif., 2006.

RMS—*See* Risk Management Solutions.

U.S. Army Corps of Engineers, "Climate Change Adaptation: Comprehensive Evaluation of Projects with Respect to Sea-Level Change," responses to Climate Change Program, web page, revised November 19, 2014. As of April 15, 2016:
http://www.corpsclimate.us/ccaceslcurves.cfm

U.S. Geological Survey, "National Elevation Dataset," 2006, accessed March 20, 2014. As of June 3, 2016:
http://ned.usgs.gov/

USGS—*See* U.S. Geological Survey.

## 3.6. Tornadoes

Tornadoes occur relatively frequently, but they usually affect small areas of land. Data about historical tornado observations can be incomplete because they require that somebody observe a tornado, which might not occur in a relatively unpopulated area or at night. Since 1950, there has been a steady increase in the number of tornadoes reported in the United States, but this rise has been driven by increases in reporting of the weakest, F0 tornadoes (Ramsdell and Rishel, 2007).

Likely because of a building's relatively small chance of being hit by a tornado, building codes do not require construction in anticipation of tornado winds. In contrast, hurricane winds, which are likely to hit a relatively large area of land, are considered in building codes.

Climate change can impact tornado frequency and severity, but there is little understanding of what changes would occur. There are two major limitations for studying convective storms. First, historical data are problematic, which leads to low confidence in measured historical trends (Kunkel et al., 2013; Seneviratne et al., 2012). As mentioned in the following section, this can lead to underestimates of the risk of tornadoes in the data we are using, but these underestimates are likely to be concentrated in less-populated regions without much infrastructure where tornadoes go undetected. The second limitation is that climate models do not have the resolution to model tornadoes. Overall, some changes in climate are likely to help tornado formation (atmospheric instability) and other changes likely to hurt (reduced vertical shear) (Seneviratne et al., 2012, p. 151). Others find that the trends tend to be favorable but statistically insignificant (Kunkel et al., 2013). Therefore, we do not model changes in tornado exposure caused by climate change.

### 3.6.1. Data Sources

Tornado wind-speed return periods were taken from the extreme wind-speed analysis conducted by Pacific Northwest National Laboratory (Ramsdell and Rishel, 2007) covering the years 1950 to 2003. While the data set covers a range of return periods, our analysis uses a 100,000-year return period.

### 3.6.2. Analysis Methods

The most-detailed study we identified on tornado likelihoods was conducted by Pacific Northwest National Laboratory (Ramsdell and Rishel, 2007) on behalf of the Nuclear Regulatory Commission. Unlike most infrastructures, nuclear facilities must consider extremely low likelihoods, such as the probability of being hit by a tornado, into their design basis.

To estimate the likelihood of a tornado strike and the associated winds, the researchers used a database of tornado strikes since 1950 kept by the National Centers for Environmental Information (formerly the National Climactic Data Center). These data provide the location, length, width, and maximum intensity. They estimated probability/wind-speed distributions for tornado strikes to either a point of land or any point within a 200-foot-wide area (to represent a large building). These estimates of distributions were made for grids laid over the United States where the cells were either 1-degree, 2-degree, or 4-degrees. Higher levels of resolution result in a more-detailed map but reduce the number of tornado observations, which makes the estimates less reliable (Figure 3.10).

Based on these findings, the Nuclear Regulatory Commission set a design basis for nuclear facilities to withstand wind speeds of 230 miles per hour in central United States, 200 miles per

**Figure 3.10.**
**Pacific Northwest National Laboratory Estimates of $10^{-5}$-Year$^{-1}$ Probability Tornado Wind Speeds, 2-Degree by 2-Degree Cells (Cells Are Overlaid on a Map of CONUS)**

| | 124 | 122 | 120 | 118 | 116 | 114 | 112 | 110 | 108 | 106 | 104 | 102 | 100 | 98 | 96 | 94 | 92 | 90 | 88 | 86 | 84 | 82 | 80 | 78 | 76 | 74 | 72 | 70 | 68 | 66 | |
|---|---|---|---|---|---|---|---|---|---|---|---|---|---|---|---|---|---|---|---|---|---|---|---|---|---|---|---|---|---|---|---|
| 47 | 0 | 0 | 0 | 0 | 0 | 0 | 0 | 0 | 78 | 73 | 119 | 97 | 123 | 105 | 105 | 101 | 142 | 120 | 79 | | | | | | | | | | 114 | | 47 |
| 45 | 0 | 0 | 0 | 78 | 0 | 0 | 0 | 0 | 0 | 0 | 88 | 116 | 130 | 134 | 135 | 144 | 155 | 146 | 138 | 133 | 103 | 136 | | | | | 135 | 0 | 75 | 97 | 45 |
| 43 | 0 | 0 | 0 | 0 | 0 | 0 | 82 | 0 | 0 | 72 | 116 | 122 | 122 | 140 | 146 | 154 | 156 | 153 | 153 | 146 | 140 | 145 | | 101 | 130 | 93 | 97 | 81 | 85 | 82 | 43 |
| 41 | 0 | 0 | 0 | 0 | 0 | 0 | 0 | 0 | 0 | 76 | 97 | 122 | 135 | 154 | 159 | 156 | 162 | 168 | 152 | 163 | 163 | 152 | 153 | 154 | 134 | 125 | 127 | 126 | | | 41 |
| 39 | 0 | 0 | 0 | 0 | 0 | 0 | 0 | 65 | 0 | 0 | 106 | 107 | 139 | 158 | 165 | 168 | 157 | 165 | 163 | 159 | 162 | 150 | 124 | 132 | 123 | 127 | 86 | | | | 39 |
| 37 | | 67 | 0 | 0 | 0 | 0 | 0 | 0 | 0 | 82 | 0 | 114 | 149 | 157 | 167 | 162 | 143 | 143 | 155 | 152 | 150 | 127 | 83 | 106 | 128 | 118 | | | | | 37 |
| 35 | | 0 | 0 | 0 | 0 | 0 | 0 | 0 | 0 | 0 | 0 | 121 | 150 | 165 | 176 | 159 | 160 | 165 | 159 | 169 | 152 | 115 | 142 | 138 | 141 | 120 | | | | | 35 |
| 33 | | | 0 | 0 | 0 | 0 | 0 | 91 | 0 | 70 | 0 | 134 | 143 | 157 | 161 | 161 | 161 | 157 | 161 | 166 | 150 | 145 | 135 | 140 | 112 | | | | | | 33 |
| 31 | | | | | 0 | 0 | 0 | 0 | 0 | 0 | 66 | 115 | 124 | 133 | 141 | 157 | 156 | 158 | 164 | 151 | 148 | 143 | 119 | 120 | | | | | | | 31 |
| 29 | | | | | | | | | | | 0 | 100 | 101 | 128 | 145 | 140 | 140 | 131 | 140 | 132 | 130 | 122 | 127 | | | | | | | | 29 |
| 27 | | | | | | | | | | | | | 130 | 106 | 118 | 72 | | | | | | 141 | 126 | | | | | | | | 27 |
| 25 | | | | | | | | | | | | | | 98 | 114 | | | | | | | 161 | 116 | | | | | | | | 25 |
| | 124 | 122 | 120 | 118 | 116 | 114 | 112 | 110 | 108 | 106 | 104 | 102 | 100 | 98 | 96 | 94 | 92 | 90 | 88 | 86 | 84 | 82 | 80 | 78 | 76 | 74 | 72 | 70 | 68 | 66 | |

SOURCE: Ramsdell and Rishel, 2007, pp. 5–7.
RAND *RR1453/1-3.10*

hour in the western Great Plains and on the East Coast, and 160 miles per hour in the western United States.

The report presents the estimate distributions in maps that show each cell and report the tornadic wind speed estimated for the map's return period. Each map has a return period of either $10^{-5}$ (once in 100,000 years), $10^{-6}$ (once in 1 million years), or $10^{-7}$ (once in 10 million years).

Additional data are presented for each cell that give the strike probability of any intensity tornado, and data that allow adjustments in likelihood to be made that account for 200-foot structures. Furthermore, the report provides a method for adjusting for other sizes of structure as well. Unfortunately, the result of these adjustments adjust only the likelihoods (e.g., a 400-foot structure in Florida would have an increase in likelihood from $10^{-5}$ to $1.059 \times 10^{-4}$ [Ramsdell and Rishel, 2007, p. A-2]), which means that cells in the adjusted maps would vary in their likelihood, which makes comparing risk difficult.

In our analysis, we consider a return period ($10^{-5}$—that is, once in 100,000 years—which is the highest likelihood available) and compare thresholds with the reported wind speeds to identify areas of high tornado hazards. Figure 3.11 shows the digitized map. White areas of the map correspond to areas where tornadoes have not been observed. Note that the observation of tornadoes is more likely in populated areas; hence areas in the sparsely populated West that have observed tornadoes tend to be in large cities like Phoenix and Salt Lake City. Therefore, the calculations are likely to underestimate tornado severity in lesser-populated areas.

**Figure 3.11**
**RAND Digitization of Pacific Northwest National Laboratory Estimates of $10^{-5}$-Year$^{-1}$ Probability Tornado Wind Speeds, 2-Degree by 2-Degree Cells**

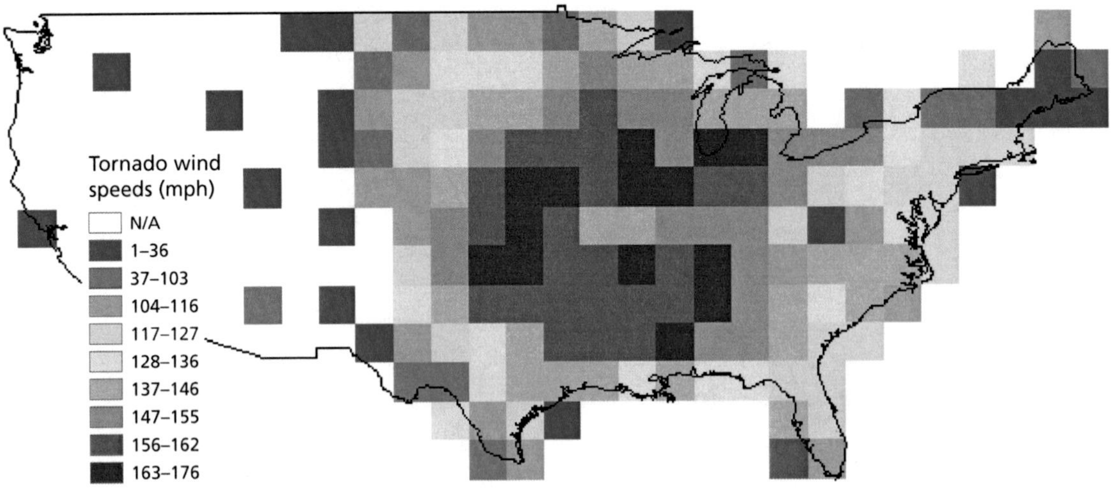


SOURCE: Ramsdell and Rishel, 2007, p. 5–7.
RAND *RR1453/1-3.11*

### 3.6.3. References

Kunkel, Kenneth E., Thomas R. Karl, Harold Brooks, James Kossin, Jay H. Lawrimore, Derek Arndt, Lance Bosart, David Changnon, Susan L. Cutter, Nolan Doesken, Kerry Emanuel, Pavel Ya. Groisman, Richard W. Katz, Thomas Knutson, James O'Brien, Christopher J. Paciorek, Thomas C. Peterson, Kelly Redmond, David Robinson, Jeff Trapp, Russell Vose, Scott Weaver, Michael Wehner, Klaus Wolter, and Donald Wuebbles, "Monitoring and Understanding Trends in Extreme Storms: State of Knowledge," *Bulletin of the American Meteorological Society*, Vol. 94, No. 4, April 2013, pp. 499–514.

Ramsdell, James V., Jr., and Jeremy P. Rishel, *Tornado Climatology of the Contiguous United States*, NUREG/CR-4461, Rev. 1, PNNL-15112, Rev. 1., prepared for Division of Risk Assessment and Special Projects, Office of Nuclear Regulatory Research, U.S. Nuclear Regulatory Commission, Washington, D.C., 2007.

Seneviratne, Sonia, Neville Nicholls, David Easterling, Clare M. Goodess, Shinjiro Kanae, James Kossin, Yali Luo, Jose Marengo, Kathleen McInnes, Mohammad Rahimi, Markus Reichstein, Asgeir Sorteberg, Carolina Vera, and Xuebin Zhang, "Changes in Climate Extremes and Their Impacts on the Natural Physical Environment," Intergovernmental Panel on Climate Change, eds., *Managing the Risks of Extreme Events and Disasters to Advance Climate Change Adaptation (SREX)*, New York: Cambridge University Press, 2012, pp. 109–230.

## 3.7. Landslides

Small landslides are relatively frequent but do not tend to have notable impacts, whereas large landslides, while rare, can have a range of impacts. In addition to affecting water supplies, fisheries, forests, and sewage disposal systems, larger landslides can cause damage to dams and transportation routes. Because landslides are highly localized events, regional or national losses associated with landslides are hard to determine. Additionally, landslides tend to coincide with other disasters, such as hurricanes and earthquakes, making it difficult to isolate the specific impacts of landslides.

Avalanches and landslides are related. We included landslides because we judge it unlikely that avalanches present a high risk for most infrastructure assets as they tend to occur in remote areas without a high density of infrastructure.

The SREX concludes with high confidence that the frequency of landslides will increase in the future because of climate change from heat, glacial retreat, permafrost degradation, as well as increased heavy rainfall (Seneviratne et al., 2012, p. 114). However, the SREX concludes there is low confidence about future locations and timing because "these depend on local geological conditions and other non-climatic factors" (Seneviratne et al., 2012, p. 114). Similarly, the data we are using to assess landslide risk are coarse and reflect general trends in susceptibility, but site-specific topography is the greatest determinant of landslide risk. Because our data do not support detailed assessment of landslide risks, we do not think they justify any climate adjustment, but should be interpreted with the realization that landslide risk is likely to increase with climate change if changes in precipitation effect areas where the elevation contours of land create a susceptibility for landslides.

### 3.7.1. Data Sources

Landslide data were drawn from the USGS Landslide Incidence and Susceptibility assessment (USGS, 2013).

### 3.7.2. Analysis Methods

Research on landslide susceptibility is limited. The most commonly cited source of country-wide landslide hazards is a map of Landslide Incidence and Susceptibility from the USGS (Radbruch-Hall et al., 1982), which has been digitized (USGS, 2013) and is included in HSIP Gold.[4]

The USGS Landslide Incidence and Susceptibility maps were created by looking at formations identified on the geologic map of the United States (King and Beikman, 1974) and estimating landslide incidence and susceptibility across the formation or groups of formations. Many of these formations are large in areas—especially east of the Rockies—so large areas have the same rating, even if local incidence or susceptibility varies. For example, the USGS incidence and susceptibility ratings do not take into account topography—a point on a flat plain would get the same rating as a point on a steep hillside if they are in the same formation.

We focus our analysis on landslide susceptibility. While the data contain incidence information, they do not provide an estimate of likelihood of a landslide for a particular time

---

[4] More-detailed research about landslide hazards that uses more-rigorous methods is available for small areas. For example, Schulz (2006) used light and detection and ranging to precisely determine where landslides had occurred in the city of Seattle.

period, only the incidence of a landslide at some point in history. Furthermore, the map is coarse, which diminishes the impact of local topography that can affect the likelihood of a landslide.

In addition to not providing a direct measure of likelihood, the available data also do not provide a direct measure of consequences either, because they do not indicate how extreme the landslides are. We assume that the consequence of a landslide would be severe if it occurred.

### 3.7.3. References

King, Philip B., and Helen M. Beikman, *Explanatory Text to Accompany the Geologic Map of the United States*, U.S. Geological Survey Professional Paper 901, Washington, D.C.: U.S. Government Printing Office, 1974.

Radbruch-Hall, Dorothy H., Roger B. Colton, William E. Davies, Ivo Lucchitta, Betty A. Skipp, and David J. Varnes, *Landslide Overview Map of the Conterminous United States*, U.S. Geological Survey Professional Paper 1183, Washington, D.C.: U.S. Geological Survey, 1982. As of April 15, 2016: http://pubs.usgs.gov/pp/p1183/pp1183.html

Schulz, William H., "Landslide Susceptibility Revealed by LIDAR Imagery and Historical Records, Seattle, Washington," *Engineering Geology*, Vol. 89, 2006, pp. 67–87.

Seneviratne, Sonia, Neville Nicholls, David Easterling, Clare M. Goodess, Shinjiro Kanae, James Kossin, Yali Luo, Jose Marengo, Kathleen McInnes, Mohammad Rahimi, Markus Reichstein, Asgeir Sorteberg, Carolina Vera, and Xuebin Zhang, "Changes in Climate Extremes and Their Impacts on the Natural Physical Environment," in Intergovernmental Panel on Climate Change, eds., *Managing the Risks of Extreme Events and Disasters to Advance Climate Change Adaptation (SREX)*, New York: Cambridge University Press 2012, pp. 109–230.

U.S. Geological Survey, "Landslide Overview Map of the Conterminous United States," web page, U.S. Department of the Interior, December 18, 2013. As of April 15, 2016: http://landslides.usgs.gov/hazards/nationalmap/

USGS—*See* U.S. Geological Survey.

CHAPTER FOUR

# Infrastructure Data Collection Process

All infrastructure data used in the completed analysis are drawn from the Homeland Infrastructure Foundation-Level Data HSIP Gold database. HSIP Gold is a geospatial data inventory jointly built by the National Geospatial Agency and DHS, and used by the Homeland Security and Homeland Defense (HLS/HD) communities.

As a first step, we mapped HSIP Gold infrastructure data to PPD-21 infrastructure *sectors* and identified a subset of infrastructure sectors on which to focus our analysis. We included a sector in our analysis when it exhibits at least a few of the following attributes:

- broader network effects within sector
- cascading effects to other sectors
- major direct property or casualty consequences.

Using the above list of criteria as guidance and erring on the side of inclusion, we arrived at the following sectors as the ones to include in our modeling effort:

- chemical industry
- communications
- energy (including nuclear power)
- transportation
- water supply and treatment.

HSIP Gold contains several infrastructure layers pertaining to each of the sectors listed above. To ensure that the model is useful without being excessively computationally intensive, we selected a subset of data layers (or subsectors) within each chosen infrastructure sector to include for each hazard in the model. Using the same guiding criteria used to pick infrastructure sectors, we first classified data layers pertaining to each of the five chosen sectors into the following four categories:

- definitely include in spatial exposure modeling
- assume exposure but do not explicitly model (redundant layer)
- likely exclude from model based on judgment (unlikely to have notable consequences)
- definitely exclude from model.

We then refined this categorization through review of our categorization with sector specific experts from the DHS Office of Infrastructure Protection. This review provided guidance on specific additional subsectors to either include or exclude. These recommendations were incorporated in the finalized list of data layers to include and exclude for each of the five sectors.

We further pruned the data to identify a subset of data elements (e.g., individual electric power plants are data elements of the electric power generation plants data layer in the energy sector) to include for each subsector. The rationale for pruning at this level is consistent with the rationale for focusing on subsets of sectors and data layers—to ensure that model results are useful and include critical elements without unduly straining computational resources. With input from sector experts, we identified a set of criteria including status (in operation vs. obsolete), capacity (any available measure of size; e.g., generation capacity for electric power generation plants), throughput (any available measure of traffic or flow), and connectivity (any available measure of node criticality; e.g., number of lines connected to a substation) with accompanying thresholds for inclusion to arrive at a final list of data elements for each chosen infrastructure subsector. For some subsectors, we chose to include all elements when either the number of total elements was small or the data layer had already undergone filtering (e.g., interstate highways).

Table 4.1 describes for each infrastructure layer the sample of infrastructure elements that are considered in this analysis. The table shows that all elements are taken into account for the majority of the infrastructure layers. For some infrastructure types, a sample of the most relevant is selected based on capacity, status, functionality or relevance. Table 4.2 lists the number of infrastructure assets by type included in this analysis.

**Table 4.1**
**Population Data Sources and Infrastructure Types Considered**

| Infrastructure Name | Types or Size of Infrastructure Considered |
|---|---|
| Electric power generation plants | Operational capacity above 10 MWs |
| Chemical industries | Annual volume sales above $100,000 |
| Energy distribution and control facilities | All |
| FAA air route traffic control centers | All |
| Natural gas import/export points | All |
| Natural gas processing plants | Active and capacity above 100 MMcfd |
| Intermodal terminal facilities | Only for trucks |
| Internet exchange points | All |
| Locks | All |
| Nuclear power plants | All |
| Nuclear fuel facilities | All |
| Petroleum, oil, and lubricants storage facilities | All active |
| U.S. coastal, Great Lakes, and inland ports | All |
| Railroad stations | All active |

**Table 4.1—Continued**

| Infrastructure Name | Types or Size of Infrastructure Considered |
|---|---|
| Refineries | All operational |
| Railroad tunnels | All active |
| Railroad yards | All |
| Substation | Number of lines equal to or more than 10 |
| Fixed-guideway transit systems stations | All |
| Airports | Only international airports |
| Canals | All |
| Road bridges | All |
| Road tunnels | All |
| Railroad bridges | All |
| Oil and natural gas pipelines | Operating status, and commodities include crude oil, natural gas, and additional oil derivatives |
| Fixed-guideway transit systems transit lines | Active or under construction; rail modes: heavy rail, light rail, and commuter rail |
| Electric power transmission lines | Extra-high or ultra-high voltage |
| Interstates | Only shield class type "1" |
| Railroad transit lines | Main rail lines, operating speed higher than 60 mph |
| Dams | All |
| Population census 2010 | All |
| Wastewater treatment plants | All |

NOTE: MW = megawatts. MMcfd = million cubic feet per day.

**Table 4.2**
**Number of Total Assets in Each Infrastructure Layer**

| Infrastructure Sector | Subsectors | No. Included in Analysis |
|---|---|---|
| Chemical industry | Chemical manufacturing facilities | 52,759 |
| Communications | Internet exchange points | 78 |
| Energy (including nuclear power) | Electric power generation plants | 4,017 |
| | Electric power substations | 870 |
| | Energy power transmission lines[a] | 208,612 |
| | Energy distribution and control facilities | 80 |
| | Natural gas import/export points | 26 |
| | Natural gas processing plants | 179 |
| | Nuclear fuel facilities | 13 |
| | Nuclear plants | 89 |
| | Oil and natural gas pipelines[a] | 1,685,806 |
| | Petroleum, oil and lubricants storage facilities | 58 |
| | Refineries | 144 |
| Transportation | Airports[a] | 180 |
| | Canals[a] | 1,186 |
| | DHS-identified railroad bridges | 114 |
| | Railroad stations | 499 |
| | Railroad transit lines[a] | 73,528 |
| | DHS-identified railroad tunnels | 30 |
| | Railroad yards | 2,211 |
| | DHS-identified road bridges and tunnels | 140 |
| | Coastal, Great Lakes, and inland ports | 22,635 |
| | FAA air route traffic control centers | 22 |
| | Fixed-guideway transit systems, stations, and lines[a] | 5,340 |
| | Intermodal terminal facilities | 3,270 |
| | Interstate highways[a] | 83,443 |
| | Locks | 219 |
| Water supply and wastewater treatment | Dams | 372 |
| | Wastewater treatment plants | 3,970 |

[a] Estimated number of infrastructure points. Point counts were estimated when the original geospatial information systems infrastructure data set type was in line or area form. In this case, centroid or distance rules were applied for point creation.

CHAPTER FIVE

# Approach to Characterizing Infrastructure Vulnerability to Hazards

The primary focus of the analysis is to characterize exposure of national infrastructure to natural disasters. As a first step, we assessed the vulnerability of each infrastructure sector to each considered hazard. If a given type of infrastructure is unlikely to be affected by a particular hazard (e.g., roads and drought), then even if this infrastructure was exposed to this hazard, there would be no impact.

We define vulnerability as the effect on an infrastructure subsector of exposure to one of the 11 natural hazards described in Chapters Two and Three—coastal floods, riverine floods, hurricane winds, extreme temperatures, droughts, wildfires, earthquakes, ice storms, tsunamis, tornados, and landslides. We found vulnerability data that can be classified into three categories—empirical, standards-based, and rooted in expert judgment. These three categories of data correspond to different levels of detail and fidelity.

Empirical vulnerability data describe the response of a given infrastructure sector to hazards of varying severity. This type of data can be seen as providing a *high* level of detail about infrastructure response to hazards. We found data of this high fidelity for only a few of the infrastructure subsectors included in this analysis (e.g., depth-damage curves—which estimate the percentage of damage to a structure with each additional foot of flood depth above first floor level—from HAZUS for a subset of energy subsectors including electric power generating plants, substations, nuclear plants, nuclear fuel facilities, and oil refineries).

In the absence of such detailed empirical data, existing design or performance standards that specify the severity of hazard that a given infrastructure asset is required to be able to withstand can provide some insight into the vulnerability of different infrastructure subsectors to different hazards. This category of data can be seen as providing a *medium* level of detail about infrastructure vulnerability because such standards are not comprehensive across sectors and across hazards of different severities. For some of the included infrastructure sectors, we found this type of standards-based vulnerability data (e.g., vulnerability of nuclear power plants to hurricane winds).

For most sectors, however, there is no documented vulnerability data, so we used expert judgment to make a binary decision (yes, if affected; no, if not affected) regarding whether an infrastructure system or subsector was likely to be affected by a hazard. This category of data can be seen as providing a *low* level of detail. Table 5.1 provides an overview of the detail level of available vulnerability data across hazards and infrastructure sectors.

One limitation of the varying fidelity of the available vulnerability data is that our analysis is necessarily limited to the lowest level of detail (the judgment approach) across infrastructures and hazards to ensure valid processing and interpretation of results.

In characterizing the exposure of infrastructures to hazards, we only count an infrastructure asset in a location as being exposed to a given hazard if we assessed that there could be a conceivable vulnerability of the infrastructure to the hazard (i.e., a "1" in Table 5.2) *and* if the hazard exists in the particular location. When assessing vulnerability, we generally considered interactions that result in direct physical damage, not service disruption or physical damage caused by subsequent reactions or cascading events. These judgments are subjective and other judgments might be appropriate for other analyses.

An important feature of this analysis is that it is possible to assess each infrastructure type's susceptibility to being affected by multiple hazards of different severities and different likelihoods. All hazards considered are classified into categories that make it possible to compare their aggregated exposure on a given infrastructure asset; Table 5.3 presents the criteria for classifying hazards into these categories.

We consider two levels of hazard severity, high and low, with high severity representing the most potentially damaging hazards, and low severity representing all potentially damaging hazards, including those also considered to be of high severity.

Additionally, we consider three likelihood periods: (a) return periods less than or equal to 100 years; (b) return periods greater than 100 years but less than or equal to 1,000 years; and (c) return periods greater than 1,000 years. Like severity, likelihood also is cumulative—when a hazard event has a return period of less than or equal to 100 years and is expected to occur in likelihood period (a), it is also included in the less likely likelihood periods, periods (b) and (c).

Allowing for variation across the two hazard severities and the three likelihood periods results in six severity-likelihood categories, represented as columns in Table 5.3. The rows represent individual hazards; each cell contains the criteria for classifying each hazard across the severity-likelihood categories (blank cells indicate an instance where the hazard does not qualify).

For example, for the earthquake hazard, high severity includes all earthquakes with PGA at or above 0.5 g, thus a 500-year return period earthquake of 0.5 g PGA falls inside of both the "High Severity; 100 yr<p≤1,000 yr" and "High Severity; p>1,000 yr" categories. Additionally, low severity also includes earthquakes of 0.5 g PGA, so this example earthquake also falls within the "Low Severity; 100 yr<p≤1,000 yr" and "Low Severity; p>1,000 yr" categories. Blank cells indicate instances where the hazard does not qualify.

**Table 5.1**
**Level of Detail of Vulnerability Data by Hazard and Infrastructure Sector**

| Hazard | Chemical Industry | Communications | Energy | Transportation | Water Supply and Treatment |
|---|---|---|---|---|---|
| Flooding (coastal and riverine) | Low | Low | Medium | Medium | High |
| Hurricane winds | Low | Low | Medium | Medium | Medium |
| Extreme temperature | Low | Low | Low | Low | Low |
| Meteorological drought | Low | Low | Low | Low | Low |
| Wildfires | Low | Low | Low | Low | Low |
| Earthquakes | Low | High | Medium | High | High |

**Table 5.1—Continued**

| Hazard | Chemical Industry | Communications | Energy | Transportation | Water Supply and Treatment |
|---|---|---|---|---|---|
| Ice storms | Low | Low | Low | Low | Low |
| Tsunamis | Low | Low | Low | Low | Low |
| Tornadoes | Low | Low | Low | Low | Low |
| Landslides | Low | Low | Low | Low | Low |

**Table 5.2.**
**Interactions Leading to Physical Infrastructure Damage Between Infrastructures and Hazards**

| Infrastructure Name | Winter Storm | Extreme Temp. | Hurricane | Landslide | Riverine | Storm Surge | Wildfire | Digital Coast SLR | Tidal Flood | Tsunami | Meteorological Drought | Quake | Tornado |
|---|---|---|---|---|---|---|---|---|---|---|---|---|---|
| Interstates | 0 | 0 | 0 | 1 | 0 | 0 | 0 | 0 | 0 | 1 | 0 | 1 | 0 |
| Road bridges | 0 | 0 | 1 | 1 | 1 | 1 | 1 | 0 | 0 | 1 | 0 | 1 | 1 |
| Road tunnels | 0 | 0 | 0 | 1 | 0 | 0 | 0 | 0 | 0 | 1 | 0 | 1 | 1 |
| Fixed-guideway transit systems transit lines | 0 | 0 | 1 | 1 | 1 | 0 | 1 | 0 | 0 | 1 | 0 | 1 | 1 |
| Electric power generation plants | 0 | 1 | 1 | 1 | 1 | 1 | 1 | 1 | 1 | 1 | 1 | 1 | 1 |
| Fixed-guideway transit systems stations | 0 | 0 | 1 | 1 | 1 | 1 | 1 | 1 | 1 | 1 | 0 | 1 | 1 |
| Railroad transit lines | 0 | 0 | 0 | 1 | 1 | 0 | 0 | 0 | 0 | 1 | 0 | 1 | 1 |
| Railroad stations | 0 | 0 | 1 | 1 | 1 | 1 | 1 | 1 | 1 | 1 | 0 | 1 | 1 |
| Railroad bridges | 0 | 0 | 1 | 1 | 1 | 1 | 1 | 0 | 0 | 1 | 0 | 1 | 1 |
| Railroad tunnels | 0 | 0 | 0 | 1 | 0 | 0 | 0 | 0 | 0 | 1 | 0 | 1 | 1 |
| Railroad yards | 0 | 0 | 0 | 1 | 1 | 1 | 1 | 1 | 1 | 1 | 0 | 1 | 0 |

**Table 5.2—Continued**

| Infrastructure Name | Winter Storm | Extreme Temp. | Hurricane | Landslide | Riverine | Storm Surge | Wildfire | Digital Coast SLR | Tidal Flood | Tsunami | Meteorological Drought | Quake | Tornado |
|---|---|---|---|---|---|---|---|---|---|---|---|---|---|
| Airports | 0 | 0 | 1 | 1 | 1 | 1 | 1 | 1 | 1 | 1 | 0 | 1 | 1 |
| FAA air route traffic control centers | 0 | 0 | 1 | 1 | 1 | 1 | 1 | 0 | 0 | 1 | 0 | 1 | 1 |
| Intermodal terminal facilities | 0 | 0 | 1 | 1 | 1 | 1 | 1 | 1 | 1 | 1 | 0 | 1 | 1 |
| U.S. coastal, Great Lakes, and inland ports | 0 | 0 | 1 | 1 | 1 | 1 | 0 | 1 | 1 | 1 | 0 | 1 | 1 |
| Canals | 0 | 0 | 0 | 1 | 0 | 0 | 0 | 0 | 0 | 1 | 0 | 1 | 1 |
| Locks | 0 | 0 | 0 | 1 | 0 | 0 | 0 | 0 | 0 | 1 | 0 | 1 | 1 |
| Internet exchange points | 0 | 0 | 1 | 1 | 1 | 1 | 1 | 1 | 1 | 1 | 0 | 1 | 1 |
| Chemical industries | 0 | 1 | 1 | 1 | 1 | 1 | 1 | 1 | 1 | 1 | 1 | 1 | 1 |
| Oil and natural gas pipelines | 0 | 0 | 0 | 1 | 0 | 0 | 0 | 0 | 0 | 1 | 0 | 1 | 1 |
| Petroleum, oil, and lubricants storage facilities | 0 | 0 | 1 | 1 | 1 | 1 | 1 | 1 | 1 | 1 | 0 | 1 | 1 |
| Refineries | 0 | 1 | 1 | 1 | 1 | 1 | 1 | 1 | 1 | 1 | 1 | 1 | 1 |

**Table 5.2—Continued**

| Infrastructure Name | Winter Storm | Extreme Temp. | Hurricane | Landslide | Riverine | Storm Surge | Wildfire | Digital Coast SLR | Tidal Flood | Tsunami | Meteorological Drought | Quake | Tornado |
|---|---|---|---|---|---|---|---|---|---|---|---|---|---|
| Natural gas import/export points | 0 | 0 | 1 | 1 | 1 | 1 | 1 | 0 | 0 | 1 | 0 | 1 | 1 |
| Natural gas processing plants | 0 | 0 | 1 | 1 | 1 | 1 | 1 | 0 | 0 | 1 | 0 | 1 | 1 |
| Energy distribution and control facilities | 0 | 1 | 1 | 1 | 1 | 1 | 1 | 1 | 1 | 1 | 0 | 1 | 1 |
| Nuclear plants | 0 | 1 | 1 | 1 | 1 | 1 | 1 | 1 | 1 | 1 | 1 | 1 | 1 |
| Nuclear fuel facilities | 0 | 1 | 1 | 1 | 1 | 1 | 1 | 1 | 1 | 1 | 0 | 1 | 1 |
| Substation | 0 | 1 | 1 | 1 | 1 | 1 | 1 | 1 | 1 | 1 | 0 | 1 | 1 |
| Electric power transmission lines | 1 | 0 | 1 | 1 | 1 | 1 | 1 | 0 | 0 | 1 | 0 | 1 | 1 |
| Dams | 0 | 0 | 0 | 1 | 0 | 0 | 0 | 0 | 0 | 1 | 0 | 1 | 1 |
| Population | 1 | 1 | 1 | 1 | 1 | 1 | 1 | 1 | 1 | 1 | 1 | 1 | 1 |
| Wastewater treatment plants | 0 | 1 | 0 | 1 | 1 | 1 | 1 | 1 | 1 | 1 | 1 | 1 | 1 |

NOTES: A cell denoting "1" indicates that the infrastructure is affected by the specified hazard. On the contrary, "0" indicates that that infrastructure is not affected by the noted hazard.

**Table 5.3**
**Criteria for Binning Natural Hazards by Intensity and Likelihood**

| Hazard Type | Low Severity p≤100 yr | Low Severity 100 yr<p≤1,000 yr | Low Severity p>1,000 yr | High Severity p≤100 yr | High Severity 100 yr<p≤1,000 yr | High Severity p>1,000 yr |
|---|---|---|---|---|---|---|
| Earthquake | | [0.1g,0.5g] and [0.5g,∞]; 500-yr return period | [0.1g,0.5g] and [0.5g,∞]; 500- and 2,500-yr return period | | [0.5g,∞]; 500-yr return period | [0.5g,∞]; 500-and 2,500-yr return period |
| Landslide | All assumed low severity; all assumed p<100 yr | All assumed low severity; all assumed p<100 yr | All assumed low severity; all assumed p<100 yr | | | |
| Meteorological drought | Consider only Q95 KBDI [400,600] and KBDI [600,800]; all assumed to be p<100 yr | Consider only Q95 KBDI [400,600] & KBDI [600,800]; all assumed to be p<100 yr | Consider only Q95 KBDI [400,600] and KBDI [600,800]; all assumed to be p<100 yr | Consider only Q95 KBDI [600,800]; all assumed to be p<100 yr return period | Consider only Q95 KBDI [600,800]; all assumed to be p<100 yr | Consider only Q95 KBDI [600,800]; all assumed to be p<100 yr |
| Wildfire | | High risk and very high risk of wildfire (fire index of 401 or higher); all assumed to be 100 yr≤p<1,000 yr | High risk and very high risk of wildfire (fire index of 401 or higher); all assumed to be 100 yr≤p<1,000 yr | | Very high risk of wildfire (fire index of 1935 or higher); all assumed to be 100 yr≤p<1,000 yr | Very high risk of wildfire (fire index of 1935 or higher); all assumed to be 100 yr≤p<1,000 yr |
| Extreme temperature | 120 degrees or higher; return periods: 20 | 120 degrees or higher; all assumed p<100 yr | 120 degrees or higher; all assumed p<100 yr | 130 degrees; all assumed p<100 yr | 130 degrees; all assumed p<100 yr | 130 degrees; all assumed p<100 yr |
| Hurricane | Category 2 wind cutoff (96 mph) and Category 4 (130 mph); use natural return period | Category 2 wind cutoff (96 mph) and Category 4 (130 mph); use natural return period | Category 2 wind cutoff (96 mph) and Category 4 (130 mph); use natural return period | Category 4 (130 mph); use natural return period | Category 4 (130 mph); use natural return period | Category 4 (130 mph); use natural return period |

**Table 5.3—Continued**

| Hazard Type | Low Severity p≤100 yr | Low Severity 100yr<p≤1,000 yr | Low Severity p>1,000 yr | High Severity p≤100 yr | High Severity 100 yr<p≤1,000 yr | High Severity p>1,000 yr |
|---|---|---|---|---|---|---|
| Riverine flooding | | | | All assumed to be high severity; all assumed p<100 yr | All assumed to be high severity; all assumed p<100 yr | All assumed to be high severity; all assumed p<100 yr |
| Tsunami | | | | | All assumed to be high severity; all assumed 500-yr return period | All assumed to be high severity; all assumed 500-yr return period |
| Ice storm | Category 4 or higher; all assumed 50-yr return period | Category 4 or higher; all assumed 50-yr return period | Category 4 or higher; all assumed 50-yr return period | Category 5 for severe; all assumed 50-yr return period | Category 5 for severe; all assumed 50-yr return period | Category 5 for severe; all assumed 50-yr return period |
| Permanent inundation | 1-ft flooding; use natural return period | 1-ft flooding; use natural return period | | 6-ft flooding; use natural return period | 6-ft flooding; use natural return period | |
| Tidal flooding | 1-ft flooding; only 20-yr return period | 1-ft flooding; only 20-yr return period | 1-ft flooding; only 20-yr return period | 6-ft flooding; only 20-yr return period | 6-ft flooding; only 20-yr return period | 6-ft flooding; only 20-yr return period |
| Storm surge flooding | 1-ft flooding; only 100-yr return period | 1-ft flooding; only 100-yr return period | 1-ft flooding; only 100-yr return period | 6-ft flooding; only 100-yr return period | 6-ft flooding; only 100-yr return period | 6-ft flooding; only 100-yr return period |
| Tornado | | | EF0 and EF3; all assumed to be p≥1000 yr | | | EF3; all assumed to be p≤1,000 yr |